Life by Chemical Evolution?

Hans R. Kricheldorf

Life by Chemical Evolution?

A Review and Evaluation of Experiments and Hypotheses

Springer

Hans R. Kricheldorf
Institute for Technical and Macromolecular Chemistry
Universität Hamburg
Hamburg, Germany

ISBN 978-3-662-70751-7 ISBN 978-3-662-70752-4 (eBook)
https://doi.org/10.1007/978-3-662-70752-4

English translation and updated edition of the original German edition "Leben durch chemische Evolution?" published by Springer Spektrum.

This book is a translation and updated edition from the German language edition: "Leben durch chemische Evolution?" by Hans R. Kricheldorf, © Springer-Verlag GmbH Deutschland, ein Teil von Springer Nature 2019. Published by Springer Spektrum Berlin, Heidelberg. All Rights Reserved.

This Springer imprint is published by the registered company Springer-Verlag GmbH, DE, part of Springer Nature.
The registered company address is: Heidelberger Platz 3, 14197 Berlin, Germany

Foreword

Approximately 90 years of research on chemical evolution and the origin of life have entailed presumably more than one thousand publications including research articles, reviews of certain sections of this topic, and books covering the entire working field. Hence it is justified, that a püotential reader of this book will ask, what a new book will contribute to knowledge and understanding of this broadly studied and discussed area of science, in as much, as this book does not primarily intend to present a simple progress report or account of the current state of art. The answer results from the observation of the author that almost all review articles and books were written in an optimistic or even over-optimistic attitude to the hypothesis that life emerged as a consequence of a chemical evolution on the early Earth. Such an attitude is, of course, a normal consequence when a scientist has spent many years or even most of his active life with research on the course of a hypothetical chemical evolution. The concept of a chemical evolution from simple gases up to the first living cell mutates from a scientific working hypothesis to a firm conviction or, in other words, to an ideology. However, when a working hypothesis is substituted by an ideology, all the phenomena known from faithful of and ideology or religion can be observed, namely reduction or even disappearance of objectivity, ignoration of experimental facts that are in contradiction to the ideology and lacking facts are preplace by wishful thinking. Hence, the primary goal of this book is to discuss the usefulness of existing experiments and their interpretation as arguments in favor of a chemical evolution.

Second, any goal resulted from the observation that author has not found any review article or book written by chemist with a broad experience in polycondensation chemistry and syntheses of oligo and polypeptides. Hence, it is the intention of the author to evaluate the hitherto known model syntheses and the products extracted from meteorites from the viewpoint of polymer chemistry. Therefore, a chapter dealing with basic principles of polycondensation and ring-opening polymerization of α-amino acid derivatives is placed ahead of the enumeration and evaluation of model syntheses and other experimental studies (i.e., Chap. 3). In this chapter, it is emphasized that the polycondensation processes discussed for the prebiotic formation of polypeptides and polynucleotides are fundamentally different from all syntheses of biopolymers in living organisms. How this significant difference may have overcome in the course of the chemical evolution has also been discussed.

Research concerning the hypothesis of a chemical evolution beginning with simple gases (CH_4, N_2, NH_3, H_2O) proceeding over formation of RNAs and proteins up to a living cell is an approach which is called "bottom up" in the Anglo-Saxon literature. Research on the composition and properties of the most primitive living cell, the complementary concept, is called "top-down" approach. This book exclusively deals with the "bottom-up" approach. A recommendable introduction and summary into the "top-down" approach also called "minimal cell theory" may be found in P. L. Luisi's book *The Emergence of Life* (2006).

If the critical comments presented in this book stimulate readers to conduct new experiments or to apply stricter scientific standards to the evaluation of experiments, this book has fulfilled its purpose.

Prof. Dr. Hans-Rytger Kricheldorf
Institute for Technical and Macromolecular Chemistry, University Hamburg,
Hamburg, Germany
kricheld@chemie.uni-hamburg.de

Acknowledgment

The author wishes to thank Prof. Saber Chatti (INRA, Sidi Thabet, Tunisia) for having drawn the formulas and Mrs. Britta Peters (Library, Dept. of Chemistry) for having retrieved copies of numerous research articles, reviews, and books.

Contents

Introduction 1

In [science] the last court of appeal is observation and experiment, but not authority.

Thomas H. Huxley

1.1 How to Define Life?

The questions, how life has emerged via chemical evolution (CE) requires at first adefinitioon of the goal of the evolution, meaning a definition of life. In this regard, the Austrian writer A. Schnitzler was extremely skeptic and wrote: "Was Leben ist, vermag kein Wissenschaftler zu sagen" (No scientist is able to define the meaning of life). Nonetheless, numerous scientists have attempted to find a definition for the phenomenon called life, and the author is not ambitious to present a new definition, but prefers to present a short summary of the various attempts made by renowned scientists and philosophers. A detailed discussion of this topic may be found in publications of the following authors: Bernal (1951, 1967), Folsom (1979), Chyba and McDonalds (1995), Rizzoti (1996), Palyi et al. (2002), Popa (2004), and Luisi (2006).

The multitude of published definitions of life alone indicates that a simple and precise definition, which satisfies the majority of scientists, is difficult to find, and it is perhaps even impossible to find it. What does simple and precise mean in this context? A simple definition was coined centuries ago in Latin for the attempt to define a definition: "Definitio fit per genus proximum and differentia specifica" (a definition results from the selection of an umbrella-term combined with a denomination of specific features which distinguish to define object from other species covered by the umbrella term). By this recipe, it is feasible to define all dead or living objects and phenomena, provided that a suitable umbrella term can be found. It is exactly the difficulty to find a suitable umbrella term which makes it hard to find a simple and precise definition of life according to the classical Latin recipe. Of

H. R. Kricheldorf, *Life by Chemical Evolution?*
https://doi.org/10.1007/978-3-662-70752-4_1

course, the term "phenomenon" may be used as umbrella term, but then it is hard to find simple and precise "differentiae specificae." Therefore, almost all attempts to define life have a more or less descriptive character, and this character depends largely on the religious, ideological, or political stance of the author. In his book "Life itself" (Crick 1983), F. Crick has formulated his aspect as follows:

> Perhaps the best way to approach the problem [definition of life] is to describe what we know of the basic processes of life, stripping off the skin of the onion until there is little or nothing left, and the to generalize what we have discovered.

At this point, it should be mentioned that numerous scientists of the nineteenth century did not consider a precise definition of life to be necessary or reasonable, because they believed that the transition from a clump of dead matter to a cluster of primitive cells would be gradual rather than erratic and excursive. Since no significant qualitative difference was seen between a clump of dead matter and a cluster of cells, the CE was understood as a continuous process. This view was, for example, outlined by the biologists and philosophers F. Rolle (1863), E. Häckel (1868), and W. Pryer (1880). Since at that time, little was known about proteins and nothing about polynucleotides such as RNA and DNA, those authors had no reason to formulate a precise chemical description of the assumed continuous transition from dead to living objects.

Eventually, it is interesting to keep the following statement of the biologist T. Dobzhansky (1991) in mind: "Nothing in biology makes sense except in the light of evolution."

With this background in mind, a compilation of definitions or descriptions of life and living organisms will now be presented below. This list of comments is not exhaustive, but representative for typical trends.

An early attempt dating back to the year 1820 originates from S. T. Coleridge, but exemplifies a rather simple, not to say superficial definition (Coleridge 1820):

> I define life [....] as a whole that is presupposed by all its parts

A surprisingly early definition/description of life can also be found in the work of the biologist and philosopher F. Engels (1894), whose view was mainly focused on proteins, since any knowledge of RNA and DNA was absent at this time (translated by the author):

> Life in its highest and lowest form is nothing else than the normal state of proteins
> ("Leben in seiner niedrigsten bis zu seiner höchsten Form ist nichts anderes als die normale Daseinsform der Eweißkörper")

Emphasis on the decisive role of self-maintaining metabolism was summarized in the term "Autopoiesis" coined by Varela et al. (1974) and Maturana and Varela (1980). This term means that a living cell is an assembly of matter which maintains its structure and function itself over the entire course of its existence. In Poerksen (2004) Maturana wrote:

> Whenever you encounter a network whose operations eventually produce itself as a result you are facing an autopoietic system. It produces itself.

After 1950 the capability of self-reproduction began dominating the definitions and descriptions of the first living cells. A characteristic first example for this trend was formulated by N. H. Horowitz in 1959 (Horowitz 1959):

> I suggest that these three properties – mutability, self-dublication and heterocatalysis – comprise a necessary and sufficient definition of living matter.

In the year 1961, Oparin published characterization of living organisms on the basis of the following six properties (Oparin 1961): (1) capability to exchange chemical compounds with the surrounding medium; (2) capability to grow; (3) capability of population growth; (4) capability of serf-reproduction; (5) capability to move; and (6) capability to achieve an activated state.

A quite different approach to the definition of life was presented by V. Kubyshkin et al. (2017):

> Life can thus be defined as a process shaped by two forms of casuality: (i) deterministic laws of chemistry and physics (Natural Laws) and (ii) a genetic program that determines all biological activities and phenomena.

Finally the aspect of evolution came to the foreground. An early example of a pertinent definition was published by C. H. Waddington (1968):

> [....] it might be claimed that the most important fact about the living objects is that they take part in the long-term process of evolution.

J. T. F. Wong formulated the following statement in 2002 (Wong 2002):

> A living system is one capable of reproduction and evolution, with a fundamental logic that demands and increasing search for performance with respect to the building blocks and arrangement of these building blocks. The search will only end when perfection or near perfection is reached.

Over the past 20 years, the s. c. NASA definition has found wide acceptance:

> Life is a self-sustained chemical system of undergoing Darwinian evolution.

After all these different attempts to define life or provide a short and precise description of living organisms, it is important for the purpose of this book to cite the Nobel Laureate J. W. Sczostak:

> Attempts to define life do not help to understand the origin of life.

It is characteristic for all the afore-mentioned definitions and descriptions of life that they concern a phenomenon which is based on chemical reactions of atoms and molecules mainly consisting of the elements carbon, hydrogen, nitrogen, and oxygen. However, a few authors were eager to define life in a much broader sense. For example, the inorganic chemist and theoretical biologist A. Cairns-Smith attributed a kind of life to the structurally variable silicate clay as discussed in more detail in Sect 2.5. Nonetheless, water plays still a decisive role as liquid reaction medium. However, both, the biochemist R. Shapiro and the physicist G. Feinberg (1990), advocated an even broader range of scenarios that share characteristic properties of the life known on Earth, but without being confined to structures and processes known from the surface of Earth. Those authors postulated that the label life should be attributed to a complete biosphere, and according to their view, a biosphere:

> ...is a highly ordered system of matter and energy characterized by complex cycles that maintain or gradually increase the order of the system through an exchange of energy with its environment.

Self-reproduction and subdivision of living organisms in species combined with Darwinian evolution are characteristic for life on Earth, but in other location of space, other mechanisms might have evolved that the order of matter and energy typical for those places is maintained over a long time or even increases. Therefore, they proposed that ordered self-repeating cycles of chemical reactions, for example, in liquid ammonia of cold planets or in the melt of silicates above 1000 °C should be considered as life according to their definition. They went even so far to postulate that cyclic processes of physical changes maintaining or increasing the given order in the absence of chemical reactions might be called a kind of life.

It is certainly useful for the progress of science when scientists widen their mental horizons and develop original ideas and concepts. However, Shapiro and Feinberg like many other scientists have taken the bait to pack more and more new aspects into an originally clear and important definition with the purpose to enhance the importance of the new aspect and the definition. This trend also exists among philosophers, and the science theoretician S. Toulmin (1968) has designed the following impressive picture:

> Definitions are like belts. The shorter they are the more elastic they have to be. A short belt reveals nothing about its wearer: by stretching it can be made fit almost anybody. A short definition applied to a heterogeneous set of examples has to be expanded and contacted, qualified and reinterpreted before it can fit every case.

These words may be enjoyed by philosophers, but from the viewpoint of scientist, Toulmin's view is not only strange, but also misleading. A definition does neither to have fit anybody nor to accommodate anybody, nor to fit a heterogeneous set of examples. A definition has to provide a description of a term or phenomenon that is as precise as possible avoiding misunderstandings or overlapping with terms that may look similar at first glance. Anyway, the author of the present book has confined his understanding of life to a phenomenon which is based on chemical reactions of molecules that mainly consist of the elements C, H, N, and O.

1.2 Why Life by Chemical Evolution?

At first, it should be clarified what the term "chemical evolution" really means. For this purpose, it is useful to cite a comment of the theoretical biologist A. G. Cairns-Smith (1982, Chap. 1):

> The established view about how life first arose is embodied in the doctrine of chemical evolution. This is not easy define as it is not a single idea, but it can be understood from the accounts such as those published by Bernal (1967), and others.
>
> At a very general level the doctrine of chemical evolution is simply:
>
> 1. That life arose from systems that obeyed the normal laws of physics and chemistry. I would not cavil with that; but there is a more restricted meaning for the term "chemical evolution" that is more or less explicit in most contemporary accounts. Almost invariably there are two other notions present which seem to me to be much more dubious, namely:
> 2. That there was a prevital progression, a natural long-term trend analogous in a limited way to biological evolution, that proceeded from atoms to small molecules to larger molecules—and finally to systems able to reproduce and evolve under natural selection; and also:
> 3. That the relevant molecules in prevital processes were broadly speaking, the kinds of molecules relevant to life now.

The question "why should life be explained as consequence of a chemical evolution?" was already answered by Oparin (1924, 1938, 1953, 1957, 1961), the father of research on chemical evolution, by the question *"how should we study a phenomenon that occurred only once in the long history of Earth?"* The biologist C. DeDuve (2002) formulated his view in his book *Life evolving, molecules mind and meaning* as follows:

> In other words, when a life was the consequence of a single creative event, regardless if caused by an individual God or an anonymous power, then scientists would be out of-work and redundant with respect to the fundamental question, how life emerged and evolved on the early Earth.

The willingness and ambition of scientist to perform research on the origin of life and the possibility of a chemical evolution are, of course, respectable and honorable property of those scientists, but it is per se no evidence that chemical evolution did exist. When the various hypotheses on the origin of life are analyzed with regard to their fundamental elements, the following picture emerges.

Hypothesis I

The exclusively scientific view of a chemical evolution purports that it started out from simple gases such as CO, CO_2, CH_4, NH_3, and H_2, and continued step by step under the influence of chemical and physical laws of nature to larger molecules from there to biopolymers and finally to the fist living cell. This fundamental hypothesis may be subdivided into two quite different concepts.

A. The version which should be discussed first is the reductionistic-deterministic variant. This version postulates that the gradual cooling of the universe beginning immediately after the Big Bang entailed an evolution of matter from the appearance of quarks via atoms and molecules up to the appearance of the first cell. Reductionistic means that the properties of an object can fully be explained by the properties of its components A, B, and C. Biophysicist and Nobel Laureate M. Eigen (1971) has postulated that emergence and properties of a self-organizing and self-reproducing super-cycle of metabolism may, in principle, be explained by the wave-mechanistic properties of all atoms involved in that process in combination with favorable environmental (experimental) conditions on the early Earth. Later (DeDuve 1991) C. Duve formulated: *"Given the suitable initial conditions the emergence of live is highly probable and governed by the laws of chemistry."*

Another exponent of this philosophy was the biophysicist H. J. Morowitz, who wrote in 1992 (p. 3) (Morowitz 1992):

"*Only if we assume that life began by deterministic processes on the planet are we fully able to pursue the understanding of life's origin within the constraints of normative science*", and on p. 13: "*We also reject the suggestion of Monod that the origin requires a series of highly improbable events ,,,,,,8…].*" And on p. 121: "*We have no reason to believe that biogenesis was not a series of chemical events subject to all the laws governing atoms and their interaction.*"

The physiologist DeDuve also called this hypothesis "deterministic," and coined the following frequently cited statement (DeDuve 2002, p. 298): "*It is self-evident that the universe was pregnant with life and the biosphere with man. Otherwise, we would not be here. Or else, our presence can be explained only by a miracle [….]*"

In 1996, DeDuve published a book entitled *Vital Dust: Life as Cosmic Imperative* (DeDuve 1996). This title underlines again his view that the emergence of life was a necessary consequence of the rich chemistry which occurred (and still occurs) in space and which also occurred on the early Earth.

The reductionistic-deterministic hypothesis is a logic and consistent scientific concept, the creditability of which can, in principle, be proven or disproven by chemical and physical research activities concerning the possibility of chemical evolution. Once again, a comment of DeDuve (1991): "*The science of the origin of life has to adopt the deterministic continuity view otherwise it would not be possible to adopt a scientific method of inquiry.*"

In Chaps. 4 and 9 and in the last Chapter (Sects. 11.3 and 11.4), the author presents his evaluation of the experimental results in the light of the above hypothesis.

B. The second version comes into play, when the reductionistic concept cannot be proven. Hypothesis I B postulates that the directed course of the CE is based on a law of nature which is a kind of overriding trend complementary to the well-known chemical and physical laws of nature. This law of nature governed the chemical evolution in direction of the final goal, the first living cell. If such a law exists, it includes the consequence that life will also emerge on other planets when the (reaction) conditions are appropriate. A unique event on Earth cannot establish a law of nature. The scientists of the nineteenth century and before attributed to all living organisms and any matter on which they were based a s.

c. *vis vitali* (Latin, meaning power of life). The anti-reductionistic scientists of the twentieth and twenty-first century say that the most primitive living cell differs from an almost identical assembly of dead biomolecules by an "added value." The first and most important scientific approach in this direction was contributed by the Nobel Prize laureate I. Prigogine (Prigogine 1955; Prigogine and Glansdorff 1971; Nicolis and Prigogine 1977; Prigogine 1980; Prigogine and Strenger 1984; Nicolis and Prigogine 1989; Prigogine and Strenger 1997). He has calculated, how ordered, s. c. "dissipative" structures may arise out of chaos. Furthermore, he has introduced time as a decisive component for a proper understanding of life, because he considered life as kinetically controlled irreversible process. Moreover, he pointed out that the phenomenon "life" represents a continuous fight against entropy based on consumption of energy (reaction enthalpy).

At this point, it should be emphasized that any concept of a directed CE has the shortcoming that it disagrees with the mainstream interpretation of the biological evolution. The modern view of the biological evolution does not include that beginning with the first cell, all living organisms "knew" that the final goal of the evolution is man. On the contrary, it is believed that each species tries to optimize its properties to enhance the chance of survival. The enormous diversification which has occurred in the course of the biological evolution may be understood as a strategy to occupy as many as possible ecological niches, and thus, to enhance the chance that at least a few species survive in the case of regional or world-wide catastrophes. He catastrophes at the end of Perm and Chalk, and their consequences illustrate this point. He increasing complexity of the living organisms which can be observed over the whole course of the evolution may be understood as part of the increasing diversification. Hence, a consistent concept of both chemical and biological evolution requires either a change of the mainstream concept of biological evolution from an undirected to a directed process, or it requires to understand and to explain the CE as an undirected process.

Hypothesis II

The second hypothesis which will be discussed here is based on the assumption of an undirected CE and is characterized by a decisive role of coincidence (contingency, chance, and hazard). Hence, it should here be nicknamed "hypothesis of the fortunate coincidence," HFC. This hypothesis is again based on a reductionistic explanation of the CE up to the level of enzymatically active proteins, high molar mass polynucleotides, and vesicle-forming fatty acids and alcohol derivatives. However, the formation of the first living cell is then ascribed to a fortunate coincidence of all favorable environmental conditions and effective assembly of all molecular components. In Sect. 2.1, calculations of probabilities of the necessary biopolymers will be discussed. The advocates of HFC argue that life is extant despite extremely low probability that all the factors required for the emergence of the first cell will coincide. Hence, a fortunate coincidence resulting in the formation of the first living cell must have existed. The luck of a gambler who makes a big fortune against all odds serves as example for the correctness of this argumentation. The reader will come across this kind of logic and argumentation again and again in the following chapters, and thus, this argumentation should be called here "a posteriori argument." It is characterized by the knowledge of extant life we have today, but the molecules forming the basis of an undirected evolution could not have known that their chemical reactions will end up about 3 billion years later with the existence of the humankind.

At this point, it is advisable to discuss the meaning of the term coincidence (contingency, chance) in more detail. For this purpose, we define a term coincidence I as a moment which determines the exact time, when a certain event happens. This moment was unpredictable, but the occurrence of this event sooner or later was unavoidable. Such a kind of coincidence belongs to Hypothesis I. In contrast, Hypothesis II is based on a coincidence which includes a creative power and initiates a new direction in the evolution of the entire universe. An event according to the definition of coincidence II can neither be proven nor disproven directly, because, as shown below in Hypothesis III, the existence of life on Earth may also have other reasons, and thus, does not prove per se that a fortunate coincidence happened. However, the question, if a fortunate coincidence of type II was at least possible as an origin of life, can, in principle, be answered by experiments demonstrating the possibility of CE. When the numerous steps required for a successful CE can be proven by model experiments and by examination of the chemistry in space, a fortunate coincidence might have had the consequence that the first living cell was formed. However, as demonstrated in Chaps. 4, 5, 6, and 7, it is absolutely unlikely that a coexistence of all components required for the formation of the first cell ever existed on the early Earth. Furthermore, it should be considered that the simultaneous presence of all necessary building blocks of the first cell does not necessarily include that transformation from the entropy driven chemistry of dead matter into the kinetically controlled biochemistry of a living cell took place. In other words, the assumption of a fortunate coincidence does not significantly differ from the belief in a creational act of a goodness. A further discussion is given in Sect. 8.3.

In this connection, the meaning of coincidence (contingency, chance) in the modern theory of biological evolution is of interest. In the second half of the twentieth century, most biologist were convinced that the evolution of a new species (including the humankind) is a consequence of chance events which in analogy to coincidence II involved a creative power which resulted in the emergence of a new species and perhaps in a new direction of the evolution. The most famous exponent of this view was the French Nobel Prize laureate Jaques Monod who published in 1970 the book *Le Hazard et la Necessite* (translated in numerous languages) in which he pointed out the emergence of life, and thus, the appearance of man on Earth was nothing but the consequence of a fortunate coincidence (Monod 1970).

The dominant role of contingent events for the progress of the biological evolution was buried after 1970 by the ongoing research in genetic and molecular biology. It was initially the Nobel Prize laureate Barbara McClintock who elaborated the concept of "jumping genes" or genetic transposition, a new concept for the genetic recombination by crossing-over during the meiosis of a cell. This theory leads to a new understanding of the biological processes responsible for the evolution of a new species. A new species emerges when the entire genome of the old species starts a reorganization process in response to the stress of changes in its environment. One point mutations in the chromosomes, e.g., by radioactivity or UV light, may be favorable for the change intended by the species and then will be tolerated. If a mutation is unfavorable, the damage will be repaired. Therefore, by chance mutations in the genome of cells play the role of an auxiliary engine or booster, but they do never determine the direction of the ongoing changes and thus not determine the direction of the evolution. In summary, the understanding of chance events in modern biology is quite different from, not to say contrary to, the definition of coincidence II and the role of a "fortunate coincidence" as origin of the first living.

Hypothesis III

A third fundamental hypothesis how life emerged on Earth is characterized by the assumption that the emergence of life was unique creative act which happened only one in the universe. This hypothesis does not negate the existence of chemical and physical laws of nature, and it does not negate a chemical evolution starting out from simple gases up to various polymers, but it negates such an evolution resulted in the formation of the first living cell. For religious people, this creative act can be attributed to the Christian God, to Allah, to any other godness or to anonymous power such as the *δυναμισ* postulated by Aristotle. Since a unique event escapes any kind of scientific research (which is confined to reproducible events belonging to a law of nature), such a creative act is considered by scientist as a non-scientific explanation of life (what is correct), but it is also labeled with a pejorative, depreciatory image (what is at least premature).

However, all the reviews and books about the origin of life read by the author have ignored that hypothesis III includes the possibility the life was created by a second Big Bang which like the first Big Bang is a unique event, but a characteristic property of our universe. Any argumentation about the origin of the first Big Bang is outside the reach of scientific methods and since the existence of time before the Big Bang cannot be proven, any causal argumentation about anything before the first Big Bang is mere speculation. The first Big Bang is simply the result of tracing back the evolution of the universe, and in analogy, a second Big Bang may be postulated as result of tracing back the biological evolution. Hence, the postulation of a second Big Bang is not less scientific than the assumption of a first Big bang and the "double big bang theory" is not less scientific than the hypothesis of a "fortunate coincidence" discussed in Hypothesis II.

Any hypothesis of unique event responsible for the creation of life is automatically confronted with the question, why this event happened on Earth about 14 Billion years after the First Big Bang. Of course, we know today that the Earth had particularly favorable conditions for a biological evolution, and only for few exoplanets such a combination of favorable conditions may be expected. Yet, even when only a few planets in our galaxy are suited for a biological evolution, one has to consider that the universe comprises millions of galaxies, so that scientific justification for a unique event on Earth is impossible to find. However, if we assume that all planets in the universe having favorable conditions for a biological evolution were birthplaces of life, we come back to hypotheses that postulate a CE as consequence of laws of nature (Hypotheses IA or IB).

For the sake of completeness, the hypothesis of *panspermia* needs discussion. This hypothesis is not listed here as Hypothesis IV, because it does not provide any explanation for the origin of life in the universe. It simply says that genes, spores, or complete cells were present everywhere in our galaxy (at least) and, when hitting the Earth, started a successful biological evolution. The first version of the panspermia theory was formulated by the Greek philosopher Anaxagoras of Klazomenai (c. 499–428 b. C.), who believed that "germs of life" are distributed over the whole universe. The number of publications containing comments on the theory of Panspermia increased in the nineteenth century, for example written by J. J. Berzelius (1979–1848), L. Pasteur (1822–1895), H. F. E. Richter (1806–1876), Lord Kelvin (1824–1907), H. L. F. von Helmholtz (1821–1894). The first complete and consistent modern theory of panspermia was published by the Swedish Nobel Prize laureate S. Arrhenius. He published in 1908 a book in German entitled *Das Werden der Welten*, in which he postulated that spores were distributed in space accelerated and driven by the light of stars, and thus played the fundamental role of "germs of life" (Arrhenius 1908).

More recently, two versions of the panspermia theory may be distinguished. First, it is assumed that "germs of life" were omnipresent in space at any time, and thus, an "infection" of the Earth was unavoidable. However, no information on the origin of the "germs of life" is provided. Modern advocates of this version are the astronomer Sir F. Hoyle (1983, 2000) and his coworker C. Wickramasinghe. In their book *Evolution from Space* (Hoyle and Wickramasinghe 1981), those authors point out that the germs of life have a cosmic origin and reached the Earth either in the form of RNA or DNA either as components of complete cells or in the form of viruses or as fragments of genes. The genes were active immediately after their arrival on Earth. They started the production of polypeptides having enzymatic activities. Hoyle and Wickramasinghe also assumed that the genetic material vagabonding in space is protected against destruction by UV light by a coating with graphite. Those authors did not explain how the 20 prAAs needed for the production of enzymes were formed on the early Earth. They apparently believed that the primordial soup necessarily contained all the building blocks required for a successful CE, and that any other kind of molecules hindering the CE were absent. The scientific character of such a panspermia theory is not far from that of a religious belief in the creation of life by a divine power.

The second version of the panspermia theory may be called directed panspermia. This version was favored by L. Orgel and by the Nobel Prize laureate F. Crick (Crick and Orgel 1973; Crick 1983). Those authors assumed that life was immediately formed in our galaxy as soon as the first planet with favorable reaction conditions for a successful CE appeared on the scene. On this planet, a high culture developed which, in turn, sent small packages containing unicellular organisms in direction of other planets to make them happy with the arrival of life and with the start of a biological evolution. Regardless which version of panspermia is taken into account, any experimental evidence is lacking so far and neither version explain how the first living cells emerged in our galaxy.

The end of this chapter is dedicated to a short history of the ideas that initiated research on the origin of life via chemical evolution. The first scientist who uttered the idea that life on Earth is the consequence of a chemical evolution was C. Darwin. In a letter addressed to the botanist J.D. Hooker (cited by Darwins son F. Darwin in 1896), C. Darwin wrote a negative comment on the hypothesis of a chemical evolution:

> It is mere rubbish thinking at the present of the origin of life; one might as well think of the origin of matter.

However, a letter written in 1871 makes clear that he made a mental U-turn in the preceding 8 years:

> It is often said that the conditions for the production of a living organism are now present which could ever have been present. But, if (and oh! What a big if!) we could conceive in some warm little pond with all sorts of ammonia and phosphoric salts, light heat and electricity etc. present that a protein compound was chemically formed ready to undergo still more complex changes, at the present day such matter would be instantly devoured or absorbed, which could not have been the case before living organisms were formed.

This letter, which is stored in an archive we of the Royal Society (London) was published for the first time in 1959, so that C. Darwin has not contributed anything to the beginning of research on CE in the first half of the twentieth century (Darwin 1959). Therefore, the Russian biochemist Alexander I. Oparin (1894–1980) is nicknamed "father of the research on CE": Oparin studied chemistry at State University of Moscow and became professor of biochemistry in 1927. His experimental work was mainly concerned with the characterization of enzymes isolated from plants. However, his international reputation as scientist is based on his first book (Oparin 1924) describing his view of the origin of life on the Earth. In 1970, he was elected president of the international society for studies of the origin of life. Since his first book was published in Russian, his work remained for a long time unknown in the western hemisphere. After 1938, an English version of his first book was published and further books or contribution to international journals appeared in English and entailed gradual dissemination of his ideas (Oparin 1938). However, meanwhile the British biochemist and genetic expert J. B. S. Haldane (Fig. 1.1) published an article entitled "The origin of Life," which became widely known before the work of Oparin was noticed (Haldane 1929). The hypotheses of both authors have one important aspect in common, namely the assumption that a chemical evolution proceeded in the primordial soup which covered most of the surface of the early Earth ("a hot dilute soup" according to Haldane). However, both authors had different concepts concerning the course of the CE and these different aspects will be discussed in more detail below. Finally, it should be noticed that both authors also had in common that they never performed experimental studies concerning their hypotheses on the origin of life.

Fig. 1.1 John Burdon S. Haldane (1892–1964) © Science Photo Library

References

Arrhenius S (1908) Das Werden der Welten. Akademische Verlagsgesellschaft, Leipiz
Bernal J (1951) The physical basis of life. Routledge & Keegan Paul, London
Bernal J (1967) The origin of life. Weidenfeld & Nicolson and World Publishing Company, London
Cairns-Smith A (1982) Genetic takeover and the mineral origin of life. Cambridge University Press, Cambridge
Chyba F, McDonalds G (1995) The origin of life in the solar system: current issues. Annu Rev Earth Planet Sci 23:215–249
Coleridge S (1820) Theory of life, p 42 (cited in Snyder A (1929) Coleridge on logic and learning)
Crick F (1983) Life itself. Simon & Schuster, New York
Crick F, Orgel L (1973) Icarus Int J Solar Syst Stud 19:341
Darwin C (1959, posthum) "Some unpublished letters", Sir Gavin de Beer (ed). Notes and Records of the Royal Society of London 14:1
Darwin F (1896) The life and letters of Charles Darwin, vol 2. Appleton & Co, New York
DeDuve C (1991) Blueprint for a cell: the nature and the origin of life. Neil Patterson, Burlington
DeDuve C (1996) Vital dust: life as cosmic imperative. Basic Books, Harper Collins
DeDuve C (2002) Life evolving, molecules, mind and meaning. Oxford University Press, Oxford
Dobzhansky T (1991) Am Biol Teach 35:125–129
Eigen M (1971) Naturwissenschaften 58:465
Engels F (1894) Herrn Eugen DDühring's Umwälzung der Wissenschaft. Dietz Berlin
Folsom C (1979) The origin of life: a warm little pond. W. H. Freemen & Co, San Francisco
Häckel E (1868) Allgemeine Anatomie der Organismen. De Gruyter, Berlin
Haldane J (1929) The origin of life. The rationalist annual 3 (reprinted in Bernal 1967, pp 242–249)
Horowitz N (1959) On defining life. In: Clark F, Synge R (eds) The origin of life on the Earth. Pergamon Press, London, pp 106–107
Hoyle F, Wickramsinghe C (2000) Leben aus dem All, Zweitausendeins, Frankfurt
Hoyle F (1983) The intelligent universe. Michael Joseph Ltd., London

Hoyle F, Wickramasinghe C (1981) Evolution from space (Evolution aus dem All. Ullstein Berlin). J. M. Dent & Sons Ltd., London
Kubyshkin V, Acevedo-Rocha C, Budisa N (2017) Biosystems 10:10
Luisi P (2006) The emergence of life. Cambridge University Press, Cambridge
Maturana H, Varela F (1980) Autopoiesis and cognition: the realization of the living. Reidel, Dordrecht
Monod J (1970) Le hazard et la necessitee. Edition De Seuil, Paris
Morowitz H (1992) Beginning of cellular life. Yale University Press, New Haven
Nicolis G, Prigogine I (1977) Self-organization in non-equilibrium systems. Wiley Interscience, New York
Nicolis G, Prigogine I (1989) Exploring complexity: an introduction. W. H. Freeman, New York
Oparin A (1924) Proiskhozhdenie Zhizny. Izd Moskovskii Rabochii, Moscow (reprinted in translation: Bernal 1967, pp 199–214)
Oparin A (1938) Origin of life. McMillan, New York
Oparin A (1953) The origin of life. Dover, New York
Oparin A (1957) The origin of life on Earth. Academic Press, New York
Oparin A (1961) Life, its nature, origin and development. Oliver and Boyd, Edinburgh
Palyi G, Zucchi C, Caglioti L (eds) (2002) Fundamentals of life. Elsevier, New York
Poerksen B (2004) The century of uncertainty. Dialogs introducing constructivism. Imprint Academy
Popa R (2004) Between necessity and probability: searching for the definition and origin of life. Springer, Berlin
Prigogine I (1955) Introduction to thermodynamic processes. Charles C. Thomas, Springfield
Prigogine I (1980) From being to becoming. W. H. Freeman, New York
Prigogine I, Glansdorff P (1971) Thermodynamic theory of structure, stability and function. John Wiley & Sons
Prigogine I, Strenger I (1984) Order out of chaos: Man's new dialogue with nature. Flamingo Publ
Prigogine I, Strenger I (1997) The end of certainty. The Free Press
Pryer W (1880) Die Hypothesen über den Ursprung des Lebens. Gebrüder Paeel, Berlin
Rizzoti M (ed) (1996) Defining life. University of Padua Press, Padua
Rolle F (1863) Ch. Darwins Lehre von der Entstehung der Arten in der Anwendung und die Schöpfungsgeschichte. J. C. Hermann, Frankfurt. M.
Shapiro R, Feinberg G (1990) Possible forms of life in the environments very different from the Earth. In: Leslie J (ed) Physical cosmology and philosophy. McMillan, New York
Toulmin S (1968) Voraussicht und Verstehen. Suhrkamp Verlag
Varela F, Maturana H, Rosch E (1974) Biosystems 5:187
Waddington C (1968) The basic idea of biology. In: Towards a theoretical biology. Allen Unwin, London
Wong J (2002) Short definition of life. In: Palyi G, Zucchi C, Cagliotti L (eds) Fundamentals of life. Elsevier, New York

Hypotheses and Birthplaces of the Chemical Evolution

2

The strongest arguments prove nothing so long as their consequences are not verified by experiment. Experimental science is the queen of science and the goal of speculation.

Roger Bacon

2.1 The Early Atmosphere, the Prebiotic Soup, and the "Protein World"

Even the first protagonists of chemical evolution were aware of the fact that the composition of the early atmosphere had a strong influence on the probability that the biomolecules needed for an evolution of the first cell would be formed. Quite early, i.e., before 1950, two schools of thought are existed. Firstly, there were the advocates of an oxidizing or at least neutral atmosphere, who considered CO_2 and CO as the primary source of carbon compounds and N_2 as the only source of nitrogen containing organic compounds. Secondly, there were advocates of a reducing atmosphere mainly containing H_2, CH_4 as source of carbon, and NH_3 as source of nitrogen compounds. The first comments on the early atmosphere came from geologists without direct connection to the hypothesis of chemical evolution. However, with the first two protagonists of chemical evolution, A. I. Oparin (1924, 1938, 1957, 1961) and J. B. S. Haldane (1929), representatives of both opposite directions of thoughts appeared on the scene.

Before world war II, the representatives of the oxidizing atmosphere were prevalent. Haldane considered CO_2 as the only carbon source of organic compounds that yielded the primordial soup which he assumed to be the basis of the chemical evolution. He also assumed that UV light provided the energy for all the chemical reactions occurring in his primordial soup. Haldane was biologist, and he adopted his vision of the early chemistry from publications of other scientists, such as Allen

H. R. Kricheldorf, *Life by Chemical Evolution?*
https://doi.org/10.1007/978-3-662-70752-4_2

(1899), Moore (1913), Moore and Webster (1913), Bequerel (1924), and Baly et al. (1921, 1922).

Oparin was fan of a reducing atmosphere similar to that known at that time from the planet Jupiter. In addition to H2 (hydrogen) compounds of several elements, above all carbon and nitrogen, seemingly formed the Jupiter atmosphere. After world war II, this view became the predominant creed stimulated by the influence of the American professor of chemistry H. C. Urey. His view had a great influence because he was awarded the Nobel Prize of 1934 for his detection of deuterium. His hobby was identification of their isotopes and molecules via their spectroscopic properties. He was particularly interested in the origin of planets and their early atmosphere. In this connection, he also hypothesized about the origin of live and the question, if live might have emerged on Earth from an early atmosphere via chemical evolution. He summarized his insights and hypotheses in an influential book entitled *The Planets: Their origin and Developments* (Urey 1952).

Since hydrogen certainly was the primary matter of the solar system, it was obvious to postulate that H_2 and reduced elements such as CH_4, NH_3, and H_2O were the main components of the early atmosphere. Urey's hypothesis was seemingly confirmed by the experiments of his student S. L. Miller who exposed such an artificial atmosphere to electric discharges and found small amounts of a few amino acids. A detailed description and evaluation of these important experiments are given in Chap. 4.

In later decades, the critique of Urey's reducing atmosphere gained strength. The first argument was based on the relatively low molar mass of Earth relative to Saturn or Jupiter. The Earth was definitely not able to bind H_2 for millions of years in this atmosphere because it escaped rapidly to the surrounding space. More important were the arguments of geologists and volcanologist who argued that the gases resulting from volcanic eruptions and emanations of geologically active surface areas must have had a decisive influence on the composition of the early atmosphere. Furthermore, the oldest minerals found on Earth (up to 3.9 billion years back) contained oxidized iron and other metal oxides. Miller was flexible and objective enough to follow this new line of argumentation and repeated after 1976 his experiments with artificial atmospheres manly consisting of CO_2 and N_2 as sources of carbon-nitrogen compounds (s. Sect. 4.1). An early comprehensive review and discussion of various aspects concerning the origin of life was published by Dyson in 1985.

Both Oparin and Haldane shared the concept that a primordial soup containing relatively high concentrations of "biomonomers" was the birthplace and origin of a chemical evolution. However, those authors did not only have opposite views of the early atmosphere, they also had different views of the chemical course of the evolution. Haldane's concept was focused on formation of biopolymers which via self-organization yielded primitive associated capable of self-reproduction. With his own words:

> The first living or half-living things were probably large molecules synthesized under the influence of suns radiation, and only capable of reproduction in the particularly favorable medium in which they originated.

The idea that the emergence of live was based on the evolution of biopolymers capable of self-reproduction (genes first) was also published before Haldane by other authors, such as Troland (1914), Lipman (1924), Morgan (1926), or Muller (1929). Other authors continued to develop this fundamental concept in various directions after 1929. The hypotheses of the s. c. "RNA-world" or "Polynucleotide-World," discussed in Sects. 2.2 and 2.3, may be considered as modern versions of this direction of thoughts.

Oparin did not share the hypothesis of "genes first," but favored the concept that formation of vesicles, models of the first primitive cells, was the key step of the early evolution. He considered the formation of "coacervate droplets," which act as precursors of vesicles. He assumed that as a first stage of this process was a formation of small droplets of water insoluble organic compounds, which adhered to each other due to secondary valences. This first stage was followed by increasing size and complexity of the droplets until compartments of vesicles were formed which enabled chemical reactions and syntheses of biomolecules in a way different from the environment. Like Haldane, Oparin did never perform experiments himself, but he believed that that the experiments of Bungenberg de Jong, published in 1932, clearly confirmed his concept (Bungenberg de Jong 1932). The compartment theories discussed below in Sects. 2.4, 2.5, 2.6, and 2.7 may be considered as modern versions of Oparin's concept. In this connection, a paper of Saladineo et al. should be mentioned, who discussed the question "Genetics first or metabolism first?" in a review article in 2012 (Saladino et al. 2012).

Oparin's concept of chemical evolution and Miller's syntheses of prAAs had the consequence that in the period 1950–1970 evolution and properties of proteins were in the foreground of research activities and discussions of chemical evolution. The name protein originates from the Greek word protos meaning the first and should indicate that proteins are the most important class of biopolymers considering their quantity in a living organism, considering their diversity and their importance as enzymes. The biochemist F. Cedrangolo (1959) wrote about the origin of proteins:

> The problem of the origin of life is in our opinion essentially the problem of the formation and of the appearance for the first time on our planet of the giant protein molecule. Once we respond satisfactorily to this problem, I believe that all the other related problems can be entirely and surely be solved.

The appreciation of the protein world reached its peak between 1965 and 1970, when Fox and Harada (Fox 1959; Fox & Dose 1977), Fox et al. (1959) reported in several publications that heating mixtures of certain prAAs yields s. c. "proteinoids," a kind of polypeptides resembling proteins (for details s. Sect. 6.1). Both chemists were of course, the most ambitious advocates of the "protein world" hypothesis. Another prominent and influential proponent of the "protein world" was the biochemist J. D. Bernal, who worked in Cambridge and published his view in a booklet entitled "The physical Basis of Live" (Bernal 1951). A more detailed description of his hypothesis followed in the year 1967 (Bernal 1967).

The enthusiasm for the "protein world" hypothesis began to shrink in the late 1960s and continuously decreased after 1970. The skepticism was grounded on two sources. First, the model experiments concerning syntheses of polypeptides from prAAs were much less successful than expected (s. Chap. 6). Second, several authors began to calculate the probability for the formation of proteins from 20 different prAAs in a primordial soup, and they also calculated the probability for the formation of a primitive organism comprising at least 200 different enzymes (considered to be the minimum for a primitive cell). The calculations outlined below were presented by the biochemist R. Shapiro (1987) in his book *Origins – A Sceptics Guide to the Creation of Live on Earth*. He demonstrated that the concept of biologically relevant protein might have been formed from a primordial soup by chance belonging to the category wishful thinking and not to the category serious scientific hypotheses.

Shapiro calculated at first the maximum of possible events that might have stimulated or enabled the formation of a primitive cell from preformed biopolymers. He assumed that the entire surface of the Earth was covered by an ocean with an average depth of 10 km an assumption which certainly overestimated the volume of the ocean by factor of at least 5. He assumed then that the reacting components needed at least a volume of 1 μm^3, so that the entire ocean offered 10^{36} such "mini-reactors." He also estimated that the chemical evolution had a duration of 1 billion years (as we known today this value is too high by a factor of 3–5). Anyway 1 billion years corresponds to 5×10^{14} min, and 1 min was the minimum time Shapiro allotted to the reproduction of a primitive cell. This figure multiplied with the afore-mentioned number of "minireactors" yielding a figure of 10^{56} as the maximum number of events that were available for the generation of a cell. Shapiro then compared this figure with the probability that a monocellular organism comprising 200 enzymes was formed. He used figures published by F. Hoyle and C. Wichramasinghe and calculated a probability of $1{:}10^{120}$ for the formation of an enzyme of 200 prAA units from 20 different prAAs. Other scientists objected that this figure is too high, because slightly different protein chains might have a similar catalytic effect and calculated a probability of 1:1020. For a monocellular organism, 2000 enzymes were considered to represent a minimum number. Multiplied with the probability of 1:1020 for each individual enzyme yields a figure of 1:1040000 as probability for the generation of a living cell by chance via chemical evolution. This figure is much, much lower than the (overestimated) figure of possible events that might have yielded a cell. This result had the consequence that F. Hoyle, C. Wichramasinghe, and numerous other scientists turned from advocates of the protein world to opponents of the concept of a evolution exclusively based on statistical events.

For comparison, the calculations of the physicist H. N. Horowitz (1986) should be presented. He considered the following concept. A sealed container filled with densely packed bacteria is heated up to a temperature where all chemical bonds are cleaved. Afterwards the container is slowly cooled, so that all chemical bonds can be formed again, but under thermodynamic control (i.e., equilibration of all species). The thermodynamically most stable molecules will be formed first and in largest quantities. Horowitz then put the question, which weight fraction of the entire mass will consist of living bacteria. The same problem may also be formulated the other way round namely with the question, how probable is the appearance of a bacterium, when the heating cooling cycle is started with only one bacterium. His calculations yielded a probability of weight fraction or generation of one living bacterium by 1>:$10^{100\,000\,000\,000}$. This ratio is even far below the unfavorable figure calculated by Shapiro and Hoyle. Regardless, how probabilities for the emergence of a living cell were calculated, the majority of scientist interested in the hypothesis of chemical evolution accepted that formation of the first cell by statistical chemical reactions in a primordial soup is not a serious scientific hypothesis.

However, a few scientist also uttered the objection that a gambler, who wins the jackpot despite a short period of time invested in gambling relative to the probability of winning, had great luck. Since life is extant on Earth, a lucky series of events resulting in formation of the first cell must have existed Luisi 2006). This line of argumentation leaves the ground of sound scientific hypotheses and ends up in wishful thinking. The belief in unique chemical wonders is not more scientific than the belief in a single creative act of God or Allah. Singularities cannot be studied and explained by scientific methods regardless of their origin and their character.

The discovery that relatively short and simple RNA chains cannot only favor replication of complementary strands of RNA, but can also possess catalytic activities, simulated after 1982 a paradigm change, so that formation of RNA chains was now considered to be the first and decisive step toward a successful chemical evolution. This hypothesis, which in the beginning of the twenty-first century became the mainstream concept of chemical evolution was labeled "RNA World" by W. Gilbert in 1986 (Gilbert 1986). Origin, advantages, and disadvantages of the "RNA World" concept will be discussed in more detail in the following section.

Nonetheless, the "Protein World" hypothesis was reanimated in the beginning of the twenty-first century by K. Ikehara (2002, 2005, 2009, 2014). That author propagated the s. c. [GADV]-theory, where G means Gly, A means Ala, D means Asp, and V means Val. Ikehara assumed that copolypeptides composed of those four prAAs were formed first, and thus were the precursors of more complex proteins, but also precursors of RNA. He assumed that those copolypeptides were formed by thermal polycondensations on the slopes of volcanoes, apparently without being informed about experiments of Fox and Harada (s. Chap. 6).

According to his hypothesis, those copolypeptides formed globular aggregates in water that were capable of self-reproduction. However, Ikehara did not explain why exactly those four prAAs played an outstanding and decisive role for the course of the molecular evolution, and he also ignored the problem of racemization (s. Sect. 10.3). Furthermore, neither any theoretical nor experimental arguments in favor of the capability of self-replication were offered. In summary, the GADV-theory is nothing but a daydream without any serious theoretical or experimental basis.

Finally, the origin of organic chemicals on the early Earth merits discussion. Two different fundamental scenarios have to be considered, first, the delivery of organic matter from space and second, the generation of organic matter on the early Earth. Both scenarios are rather complementary than alternatives. As commented in Chap. 9, numerous scientists consider the delivery of chemicals from space via comets, meteorites and, primarily, micrometeorites an important source of organic chemicals. A terrestrial origin of amino acids, imino acids, and carboxylic acids from simple gases such as CO_2, Co, N_2, NH_3, and H_2 is discussed in Chap. 4, and model reactions concerning a terrestrial origin of saccharides, nucleobases, and nucleosides are described in Chap. 5. The formation of simple organic chemicals under the influence of FeS and NiS catalysts in the neighborhood of "black smokers" is mentioned in Sect. 2.6. Speculations bare of experimental basis concerning the production of organics in clay compartments is outlined in Sect. 2.5. Hydrothermal systems in cracks of the continental crust in another speculative scenario are discussed in Sect. 2.8. Yet, it has recently be shown (Matreux et al. 2024) on the basis of experiments that the heat flow in rock cracks may have concentrated and activated organic chemicals.

Also quite recently (Peters et al. 2023) a new experimental study of the Trapp group has demonstrated that tiny iron particles may have played an active role as catalysts and reducing agents for the production of organics from CO_2. Those authors wrote:

> We experimentally demonstrate that that iron rich meteoric particles and volcanic particles activate and catalyze the fixation of CO_2, yielding the key precursors of live building blocks. The catalysis is robust and produces selectively aldehydes, alcohols and hydrocarbons independent of the redox state of the environment. It is facilitated by common minerals and tolerates a broad range of the early planetary conditions. (150–300 °C, <10–50 bar, wet or dry climate). We find that up to 6×10^4 kg/year of prebiotic organics could have been synthesized by this planetary scale process from the atmospheric CO_2 on Hadean earth.

What is missing for an ideal scenario are processes yielding nitrogen containing biochemicals.

2.2 The RNA World

The hypothesis that at an early stage of the chemical evolution oligo- and poly(ribonucleotide)s were formed capable of self-replication was first formulated by C. Woese in 1967. Almost at the same time (1968) L. E. Orgel published a treatise about the origin of the genetic code, in which an early formation of RNA and DNA preceded the formation of proteins and other biopolymers. As already mentioned, the term "RNA World" was coined about 20 years later by W. Gilbert. The concept of a RNA world first was further elaborated by numerous research groups in the following years: Gesteland and Atkins (1993), Joyce (2002), Joyce and Orgel (1993, 2006), Orgel (2004), Yarus (2006), Chen and Szostak (2006), Szostak (2012), Schrum et al. (2010), Robertson and Joyce (2012), Blain and Szostak (2014), Müller (2022). In contrast to the papers of Woese (1967) and Orgel (1968), these publications included new experimental results.

Until 1982, the discussion of the "RNA World" hypothesis was a debate exclusively based on theoretical arguments. Yet, in the following years, numerous publications appeared that reported on catalytic activities of RNA chains of different origins: Czech (1987, 1993), Kruger et al. (1982), Zaug et al. (1983), Zaug and Czech (1986), Guerrier-Takada and Altman (1984), Altman et al. (1986), and Gandiner et al. (1985). These findings were of paramount importance not only for the hypothesis of chemical evolution but for the entire biochemistry and marked a paradigm change, The Czech group observed that segments of the ribosomal RNA of *Tetrahymena thermophila* catalyzed the self-splitting of ribosomal RNA. Furthermore, they observed that segments of ribosomal RNA were capable of splitting or synthesis of RNA chains having complementary base sequences in in vitro experiments. The Altman-Guerrier-Takada group studied the RNAse of the bacterium *Escherichia coli*, which is responsible for the in vivo synthesis of transfer RNA. They found that this enzyme may cause splitting of transfer RNA precursors in in vitro experiments and in absence of any protein. With these results, the paradigm that only proteins can act as biocatalysts was annihilated. Hence, these results were a revolution and sensation for biochemists and molecular biologists. The previous paradigm that only proteins may act as biocatalysts (enzymes) was a major reason why the "Protein World" had looked attractive at first glance.

However, the enthusiasm for the RNA world was initially by the fact that no experiments were in sight that explained the spontaneous formation of oligo- and poly(nucleotide)s under prebiotic conditions. The comment of Cairns-Smith (1986) concerning this aspect reads as follows:

> There have indeed been many interesting and detailed experiments in this area. But the importance of this work lies to my mind not in demonstrating how nucleotides could be formed on the primitive Earth, but in precisely the opposite: these experiments allow us to see in much greater detail than would otherwise had been possible, just why prevital nucleic acids are highly inplausible.

In the following text, Cairns-Smith enumerates 18 arguments against a prebiotic formation of poly(nucleotide)s. In this list, the phenomenon of enantiomeric cross-inhibition discovered by Joyce et al. in 1984 is not included yet (Joyce et al. 1984). Cross-inhibition means that in an enzymatic synthesis of an RNA controlled by a preformed oligoRNA (which serves as template), the presence of a L-ribonucleotide in the reaction mixture stops the chain growth. This finding looked at first glance like a deadly argument against the RNA world, because it is certainly unlikely that the primordial soup almost exclusively contained D-saccharides and L-prAAs.

In the year 1993, a conference dedicated to the "RNA World" was organized in Cold Spring Harbor (Long Island, USA), the contributions of which were published by Gesteland and Atkins in the same year (Gesteland and Atkins 1993). Twenty three oral presentations concerned properties of various types of RNA, but not even one single contribution reported on experiments dealing with the prebiotic synthesis of RNAs. In a theoretical treatise entitled "Catalysis and prebiotic RNA Synthesis," the chemist J. Ferris (1993) wrote in the introduction:

> There is a growing consensus that the RNA World did not evolve directly from molecules formed by prebiotic processes (Orgel 1986; Joyce et al. 1987; Joyce 1989). The reasoning is that the chemical processes involved in the formation of RNA oligomers are too complex

> to have occurred on the primitive Earth. It is suggested that the original life was simpler and that it evolved to the RNA World. Indeed, there is evidence to support the hypothesis that it is unlikely that the RNA World emerged from simple prebiotic molecules.

This situation had two consequences. On the one hand, new hypotheses were elaborated which explain the emergence of RNAs as consequence of simpler and chemically more stable precursor polymers. Those precursors were considered to have "invented" the principle of self-replication and thus the basis for the evolution and optimization of RNAs. Potential precursor systems supporting this hypothesis are discussed in more detail in Sects. 2.3 and 2.4. On the other hand, another group of scientists maintained their belief in a direct evolution of RNAs from simple monomeric building blocks, such as nucleobases, D-ribose, and activated phosphorylating agents. This concept was supported by numerous experimental results, contributed by numerous research groups over the past 25 years. At this point, a report of Unrau and Bartel (1998) is worth mentioning. Those authors detected that certain types of RNA are capable to catalyze various biochemical reactions in particular the synthesis of nucleotides, the building blocks of RNA. Further important results were published by the groups of M. W. Prowner and J. D. Sutherland, who found reaction pathways allowing for the formation of nucleotides under alleged prebiotic conditions. This work and other reports about synthesis of nucleosides and nucleotides under prebiotic conditions are discussed in more detail in Chaps. 5 and 7. Finally, the results of Johnston et al. (2001) and Lincoln and Joyce (2009) merit mentioning. Those authors found that certain RNAs are capable to catalyze their self-replication (s. Sect. 8.2). All these new results represent a significant progress for the hypothesis of the RNA world and in the year 2006 Joyce and L. E Orgel published a review article with a considerably more positive view than 20 years before (Joyce and Orgel 2006):

> After considering the emergence of self-replicating ribozymes from pools of random polynucleotides and recognizing the difficulties that must have been overcome for RNA replication to occur, in a realistic prebiotic soup, we now face the challenge of constructing a realistic origin of the RNA World [....]. One can sketch out a logic order of events, beginning with prebiotic chemistry and ending with DNA/protein-based life. However, it must be said that the details of this process remain obscure and are not likely to be known in the near future. The presumed RNAWorld should be viewed as a milestone, a plateau, in the early history of life on Earth. So too, the concept of RNA World has been a milestone in the scientific study of life's origin. Although this concept does not explain how life originated, it has helped to guide scientific thinking and has served to focus experimental efforts. Further progress will depend primarily on new experimental results, as chemist, biochemist and molecular biologists work together to address problems concerning molecular replication, ribozyme enzymology and RNA based cellular processes.

The results achieved by the research groups of G. F. Joyce, J. W. Szostack, D. P. Bartel, P. Umrau, P. Holliger, and other scientists after 2005 have considerably raised the credibility of the world hypothesis. Their experiments and the concept of "evolution in the test tube" will be discussed in Sect. 8.2 in more detail.

For a final evaluation of the RNA world hypothesis, it is also important to consider that the prebiotic formation of nucleotides and RNAs is not the only challenge. An accumulation of RNA in a prebiotic soup is absolutely unrealistic. This shortcoming is particularly important in connection with hypotheses that require high temperatures. Such hypotheses concern, for example, the assumption that "hot vents" on the ground of the oceans, might be the birthplaces of the chemical evolution. The s.c. "black smokers" (Sect. 2.5) or the "white smokers" (Sect. 2.6), which have temperatures in the range of 70–90 °C along with acidic or alkaline pH values, are extremely unfavorable for accumulation of RNAs. One might object that primitive organism exist in the close neighborhood of "black smokers" despite temperatures up to 90 °C. However, the survival of these organisms requires the existence of efficient repair mechanisms which heal any damages RNA and DNA may suffer from incorrect replication or from hydrolysis. The hypothesis of Biebricher and Trinks (Sect. 2.8), which favors the ice/water interface as cradle of the chemical evolution, has a big advantage with regard to the stability of RNAs and DNA.

2.3 RNA Precursors and the "Poly(nucleoamide) (PNA) World"

The question, what kind of simpler and more stable copolymers might have invented the process of replication and preceded formation of RNAs has attracted increasing interest of various scientists after 1986. In the year 1987, Joyce, Schwartz, Miller, and Orgel proposed a class of polymers having a polyphosphate structure, but D-ribose was replaced by simpler 1,3-dios (Scheme 2.1) Two diols were achiral (C and D, Scheme 2.1), whereas the other diols were chiral (Joyce et al. 1987). The afore-mentioned authors considered that formation (and reactions) of such achiral diols were much more likely than that of D-ribose in an early state of the evolution. Furthermore, they considered that formation of polyphosphates based on achiral diols was not hindered by enantiomeric cross-inhibition. The stereochemical properties of these polyphosphates may be described in analogy to those of monosubstituted polyolefins, such

Scheme 2.1 Ribonucleosides (**a**) and analogs of non-cyclic diols or triols (**b**), (**c**), (**d**)

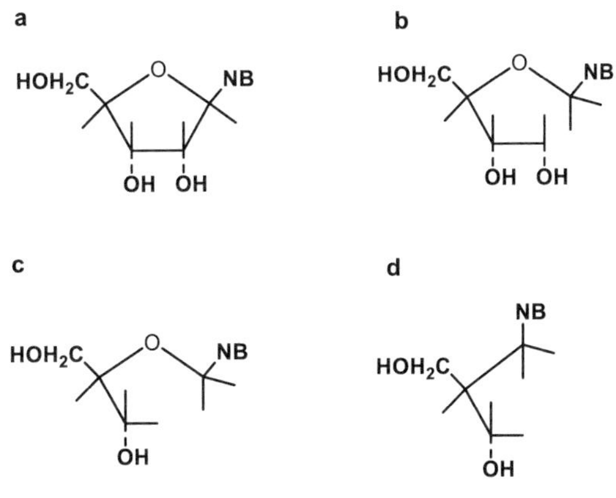

as polystyrene. For example, isotactic polystyrene, is a meso compound, but the individual building blocks turn chiral, when comonomers are incorporated in a random or alternating sequence. Isotactic polyphosphates of the diols C or D represent a similar scenartio. In corporation of two or more different nucleobases entails chirality. Joyce et al. also discussed formation of helical secondary structures and speculated about the gradual transition to the RNA world:

> If RNA is such an unlikely candidate for use as a first generation material, the why should it be any more suitable at a later stage of the evolution? Presumably many of the problems associated with the replication of the RNA were solved by development of appropriate catalytic functions in an anchestrial genetic polymer. [….] Apparently the physical and chemical properties of RNA were advantageous for its primitive role as information carrier and catalyst, and eventually allowed it to usurp the function of some precursor genetic materials during the early history of life on Earth.

Unfortunately, this visionary hypothesis was not accompanied or followed by experiments supporting it. Even the synthesis of the monomers designed in Scheme 2.1 was studied.

A quite similar approach was published by Zhang et al. (2005), who used substituted 1,2-diols as building blocks of polyphosphates (a in Scheme 2.2). They successfully synthesized monomers and oligomers containing different nucleobases, and they were able to demonstrate that these oligomers were (unexpectedly) capable

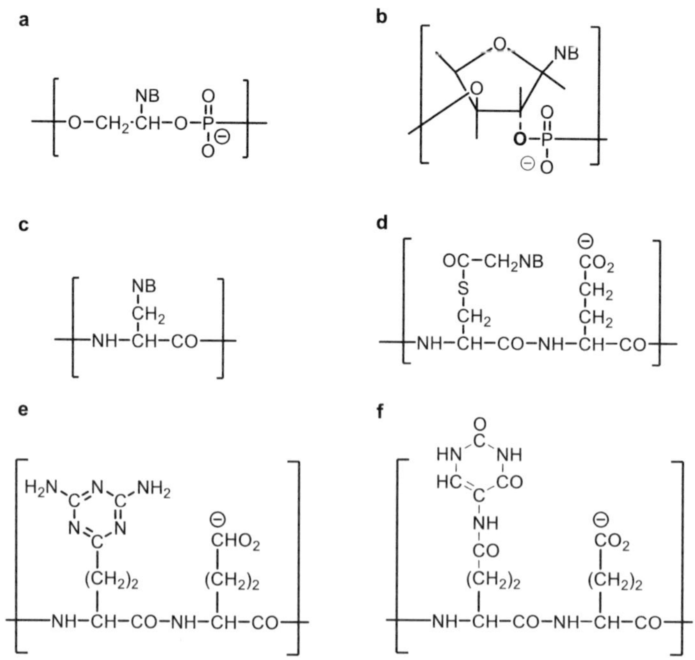

Scheme 2.2 Hypothetical examples of primitive precursors of ribonucleic acids (**a–f**)

of forming rather stable double helices. How these polyphosphates were formed under prebiotic conditions remained an unanswered question.

Another type of RNA precursor based on a polyphosphate backbone, the synthesis of which was much more likely under prebiotic conditions, was proposed by Eschenmoser and coworkers (Schöning et al. 2000). Those authors used the C4-saccharide threose (b in Scheme 2.2) as building block of the polyphosphate backbone, and despite one carbon less compared with ribose, the resulting pseudo-poly(nucleotide)s were capable of forming stable double helices with RNA via the Watson-Crick base pairing process. This success is particularly interesting because formation of C_4 saccharides, above all threose, under prebiotic conditions is much more likely than formation of C_5-saccharides with the rather unstable d-ribose. Threose is the direct reaction product of glycolaldehyde, the dimer of formaldehyde, and C_4-saccharides were identified under the condensation products of formaldehyde isolated from meteorites (s. Chap. 9).

Three research groups studied the question, if polypeptides bearing nucleobases as substituents (s. c. polynucleopeptides, PNPs) might have played the role of RNA precursors. Diedrichsen (1996, 1997) synthesized polyalanines having nucleobases as substituents in ß-position (c in Scheme 2.2). The most conspicuous result of Diedrichsen studies was the finding that formation of double helices via base pairing was only possible when the polyalanines consisted of an alternating sequence of L- and D-units. Ura et al. (2009) synthesized homochiral copolypeptides of Glu and cysteine units, the SH group of which were esterified with carboxymethyl derivatives of nucleobases (d in Scheme 2.2). However, this class of copolypeptides is certainly a rather unlikely candidate as prebiotic precursor of RNA, because esters of mercaptans are highly reactive (as known from coenzyme A) and react easily with amino groups. Therefore, their formation was impossible in the presence of ammonia, and when they were present, they had hindered the polycondensation of dipeptides and higher oligopeptides.

Eschenmoser and coworkers (Mittapali et al. 2007a, b) synthesized homochiral copolypeptides consisting of Asp and amino acids bearing a nucleobase or diaminotriazine group in the side chain (e and f in Scheme 2.2). Here the capability of base pairing was again considered to be decisive for a potential role of RNA precursors. In addition to the unanswered question of the prebiotic formation (s. point 4 in the discussion of the PNA world below), all concepts based on an early formation of polypeptides deserve the following critique. The PNP concept postulates that copolypeptides having nucleobases in their side chain are the first biopolymers formed in an early stage of the chemical evolution. They were assumed to have invented self-replication and subsequently invented RNA as a more efficient successor and alternative. Afterwards the PNPs have somehow vanished and the RNAs have, in turn, invented the synthesis of new copolypeptides and finally the synthesis of proteins. Frankly speaking, this highly speculative sequence of events looks to the author like a mental *salto mortale*.

In their review article, Robertson and Joyce (2012) have written the following comment on RNA precursor polymers:

> The hypothesis of genetic material completely different from nucleic acids has one enormous advantage – it opens the possibility of using very simple easily synthesized prebiotic monomers in place of nucleosides. However, it also raises the new and critical question Which prebiotic monomers are plausible candidates as components of replicating systems? Why would an initial genetic system invent nucleic acids once it has evolved sufficient synthetic know-how to generate molecules as complex as nucleotides?

The first sentence deserves critique. For reasons discussed in Chaps. 4, 6, and Sect. 10.3, a prebiotic synthesis of well-defined copolypeptides is as difficult to explain as a prebiotic synthesis of RNAs, above all, if homochiral polypeptides or polypeptides with alternating stereosequence are required. However, the author fully agrees with the second sentence of the Robertson-Joyce comment.

After 1986 one more class of potential RNA precursors found increasing interest as precursor polymers of RNA, namely poly(nucleoamide)s, PNAs, which are frequently, but erroneously also called poly(nucleopeptide)s. This confusion originates from the fact that the building blocks are α-amino acids, but these amino acids have an additional amino group in ω-position, and because this ω-amino group is involved in the formation of the polymer backbone, the resulting polymers must be classified as polyamides (s. Scheme 2.3). The lengths of these building blocks and the position of the nucleobases in the side chains were selected so that formation of double helices via base pairing with RNA was possible. This property should explain why these PNAs were capable to invent the formation of RNAs.

This concept was first formulated by Westheimer in 1987, but without any experimental results (Westheimer 1987). After 1992, this concept was adopted by Nielsen (1993):

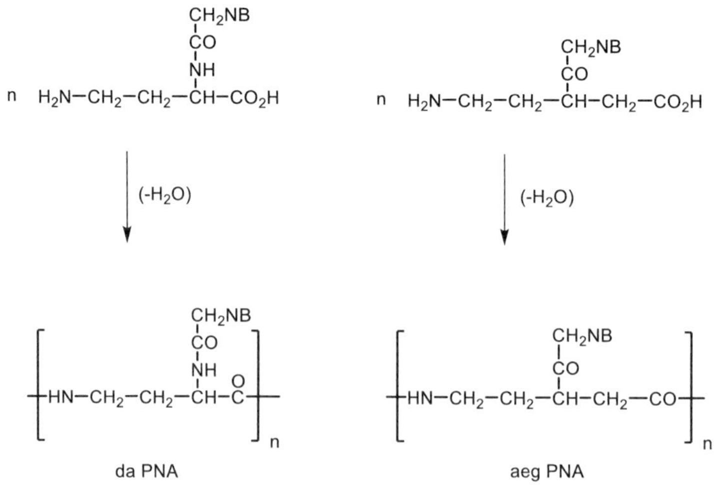

Scheme 2.3 Hypothetical examples of building blocks of polynucleoamides that have been postulated as chemically more stable precursors of ribonucleic acid

> Due to the chemical fragility of RNA it is, however, unlikely that prebiotic life could have relied on RNA [....]. We have recently described a novel DNA analog, PNA (peptide nucleic acid), which might be relevant for this discussion [....] and we have found that PNA binds to oligo(deoxy ribonucleotides obeying the Watson-Crick base-pairing rules, i.e. A – T and G – C base pairs are highly preferred. Thus, in a chemical sense, but not in a functional sense, PNA bridges the gap between proteins and nucleic acids, and the results obtained with PNA clearly show that molecules with the potential of carrying genetic information are not required to contain either phosphates or sugars, but could be peptides.

In the following years, theoretical and experimental aspects of the PNA hypothesis were worked out by Nielsen and various colleagues and coworkers in much detail (Egholm et al. 1992, 1993; Böhler et al. 1995; Schmidt et al. 1997; Koppitz et al. 1998, Kozlow et al. 2000). The base pairing and energetic interactions between various PNAS and RNA or DNA were in the focus of those studies. In their review article, Robertson and Joyce (2012) made a positive comment: "Thus it seems that a transition from a PNA World to an RNA World is possible."

However, the most important question, how were these PNAs formed under prebiotic conditions was not answered. Nielsen and partners did neither publish a theoretical concept nor any experimental results. Hence, it is of extraordinary interest to see, how the Nielsen group synthesized their PNAs. As outlined in Scheme 2.4, the first synthetic step was fixation of a BOC protecting group at the w-amino group of *N*-(aminoethyl)glycine, followed by the acylation of the α-amino group with the pentafluorophenyl ester of carboxymethyl thymidine.

The double-substituted diamino acid was ten activated as pentafluorophenyl ester, and oligoamides were synthesized in a stepwise manner via the Merifield

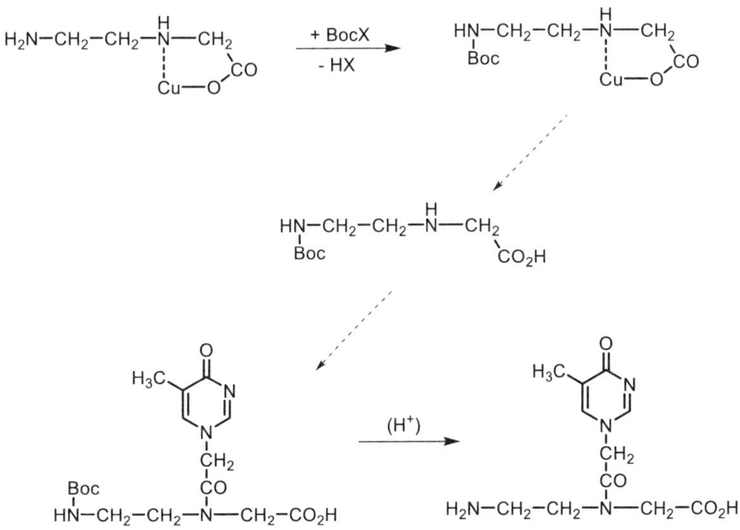

Scheme 2.4 Targeted substitution of *N*-(2-aminoethyl)glycine at the α-amino group using the Boc-protecting group method

method. The authors Egholm et al. (1993) relied on classical protecting group techniques elaborated for peptide syntheses and were not able to develop a synthetic approach according to their own concept of prebiotic evolution.

Several years later, the chemist U. Meierhenrich confessed to be a fan of the PNA world hypothesis (Meierhenrich et al. 2004). Starting out from the extraction of tiny quantities of α,ω-diamino carboxylic acids from the Murchison meteorite (S. Sect. 9.4), he designed a concept based on the assumption that α,ω-diamino carboxylic acids delivered from space to the early Earth stimulated the chemical evolution of oligomers and polymers capable of self-reproduction. With his own words:

> The delivery of organic compounds by meteorites, interstellar dust particles or comets to the early Earth is one mechanism thought to have triggered the appearance of life on Earth [....]. In the case of evolution of genetic material is it widely accepted that todays "RNA-Protein-World" was preceded by a prebiotic system, in which RNA oligomers functioned both as genetic material and as enzyme-like catalysts. In recent years oligonucleotide systems with properties resembling RNA have been studied intensively, and interesting candidates for a potential predecessor of RNA were found. Among these candidates a peptide nucleic acid molecule (PNA) is an uncharged analog of a standard nucleic acid in which the sugar phosphate backbone is replace by a backbone hold together by amide bonds. The backbone can either be composed by *N*-(2-aminoethyl) glycine (aeg) or its structural analog leading to aegPNA molecules or by a α,ω-diamino acid (da) leading to daPNA [Scheme 2.3]. The nucleobases adenine, uracil, guanine and cytosine are attached via spacers to the obtained PNA structures.

For the following reasons, the PNA concept is totally unrealistic:

1. Nucleobases having covalent spacers were never detected in meteorites.
2. Nelson et al. (2000) were able to synthesize small amounts of nucleobases having carboxymethylene spacers under alleged prebiotic conditions. However, when such syntheses were efficient under prebiotic conditions, they had hindered efficient syntheses of normal nucleosides.
3. The formation of monomers with structure A in Scheme 2.2 is in contradiction to all what is known about the chemistry of Orn, Lys or α,ω-diamino carboxylic acids. The w-amino group is more nucleophilic than the α-amino group, any electrophilic substitution will primarily involve the ω-amino group, and a mixture of products will be formed. Complexation with metal ions will involve the carboxylic group together with α-amino group and reduce its nucleophilicity even more. Divalent cations such as Mg^{2+} or Ca^{2+} were present in the ocean at any time. For preparative purposes (selective substitution of ω-amino groups), the organic chemists use Cu^{2+} ions, because these ions can easily be removed by means of H_2S after successful ω-substitution (Scheme 2.4).
4. The model syntheses of prAAs under alleged prebiotic conditions commented in Chap. 4 have shown that the formation of the desired prAAs was accompanied by formation of numerous non-protein amino acids, which have prevented successful prebitotic formation of proteins. The same results were obtained by extraction of amino acids from meteorites (Chap. 9). In other words, even the desired α,ω-diamino carboxylic acids were available on the early Earth, a simple

homopolycondensation was absolutely unlikely. Incorporation of α- or ß-amino acids in the PNA backbone would have suppressed its capability to form double helices with RNAs. This critique also holds for polycondensations of all other nucleoamino acids discussed above.

5. The condensation of N^{α}-substituted ω-amino acids in solution mainly proceeds by cyclization, so that lactams or cyclic oligoamides are formed (Eq. 1 in Scheme 2.5). This critique was already uttered by Orgel in 1998, whereupon he negated its own positive view of the RNA precursor world published together with Joyce in (Orgel 1998; Joyce et al. 1987). Miller maintained his positive view of the precursor hypothesis even in 1997.
6. The alternative approach, polycondensation of unsubstituted α,ω-diamino carboxylic acids first, followed by substitution of the remaining free α-amino groups, is again unrealistic. Polycondensations of trifunctional monomers yield hyperbranched polymers (Eq. 2 in Scheme 2.5) and not perfectly linear chains. Polycondensations of trifunctional monomers were studied by numerous research groups, and only one review article should be mentioned (Kricheldorf 2014).

Each one of the arguments 3–6 suffices to "kill" the PNA hypothesis. In summary, it may be said that the PNA world hypothesis is characterized by a strong belief of its fans in chemical wonders. The PNA hypothesis is an excellent example for demonstration of the difference between non-scientific and anti-scientific hypotheses (or theories). A non-scientific hypothesis, such as an emergence of the first living cell as a unique lucky event, can never be proven by scientific methods, but does not violate established laws of nature. However, the concept of a PNA world is in contradiction to well-known laws of organic chemistry, and thus, meets the requirements for classification as anti-scientific hypothesis.

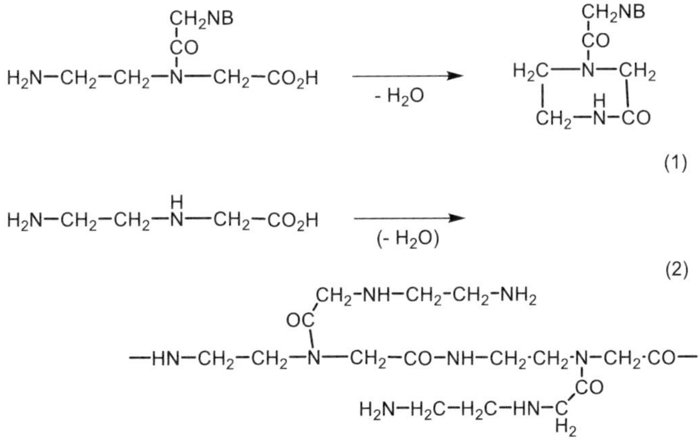

Scheme 2.5 Condensation reactions that oppose a prebiotic formation of poly(nucleoamides)

2.4 Compartment Theories

What do “compartment theories” mean? The author summarizes under the term compartment theories all concepts that hypothesize the existence of compartments regardless if their structure had a decisive influence on the course of the chemical evolution or if they were a basic requirement for the emergence of the chemical evolution. Organic compartments were also seen as precursors of the first living cells. These compartments were assumed to combine at least two important properties:

1. They enabled concentration of organic chemicals relative to the surrounding prebiotic soup and thus favored bimolecular reactions, above all, polycondensations.
2. They enabled (for example by catalytic effects) an evolution of chemical reactions which did not occur outside the compartments. In other words, they developed a special metabolism.

The father of CE research, I. A. Oparin, designed the first compartment theory. He assumed that s. c. coacervate droplets served as a first form of organic compartments which favored concentration of prAAs that finally enabled formation of the first primitive proteins. As mentioned above, he did not perform experiments himself and deduced his concept from experiments of other research groups. He described his view in a contribution of the “Report on the International Symposium *The origin of Life on the Earth*” held in Moscow in 1957 as follows:

> In our view the type of organization peculiar to life should originate only as a result of the evolution of a multimolecular organic system, separated from its environment by a distinct boundary, but constantly interacting with this environment in the manner of “open systems”. Since, as evidenced by a number of features present-day protoplasm possesses a coacervate structure, the mentioned systems, which represent the starting point for the evolution leading to the origin of life, could have been coacervate drops. But in contrast to the coacervates artificially obtained by laboratory methods these drops had to possess a certain pattern of reactions interlinked in time as characteristic to open systems. Moreover, in the course of their evolution the initial systems must gradually have become more complex and elaborate both in space (in regard to their structure) and in time (i.e. in relation to their reaction pattern).

He also discussed and studied (in cooperation with other scientists) the evolution of complex coacervates which combined proteins, lipids, and polynucleotides:

> Since initial systems of this type were bound to be destroyed at early stages of the development of life owing to natural selection, and could not have up to the present time under natural conditions, we can put the suggested processes of evolution to the test either by gradually increasing the complexity of artificially obtained systems, or by destroying to a certain extent natural protoplasmic structures and making a study of the fragments thus obtained. It is in these two directions that our experiments have been carried out; of course they can only be regarded as preliminary ones.
>
> To obtain the former aim, we chose, in collaboration with T. Evreinova, T. Shubert, M. Nestyuk, and G. Larionova artificially prepared coacervate drops. To make their organi-

> zation approach more closely the organization of protoplasm it was necessary, in the first place, to induce in these drops chemical transformation disturbing their static type of stability. This can most readily be achieved by incorporation, into the coacervate drops, of enzymes which catalyze transformation of substances forming part of these drops.

Oparin's coacervate hypothesis had two weak points. First, he believed that polypeptides and RNA-like biopolymers were formed in the primordial soup a hypothesis which lacks experimental evidence even in the year 2022.

Around 1960, the research group of S. W. Fox elaborated a new compartment theory on the basis of synthetic proteinoids, the synthesis, and structure of which is discussed in Chap. 6.

In the following years, these phenomena (i.e., proteinoids) were studied by other research groups and their reproducibility was, in principle, confirmed. Young (1965) found that even cooling of a saturated proteinoid solution from 25 °C down to 0 °C suffices generates microspheres. However, Young also criticized Fox's hypothesis of microsphere division in the sense of proliferation:

> It is quite obvious that there is a tremendous diversity of size and morphology in these microspheres, and the structure is dependent on experimental variations. For example, changes in the salt concentration or the ionic species give corresponding variations in the resultant microspheres. The pH of the medium, the temperature and pressure used, as well as the treatment of the microspheres after their formation all have an effect on microsphere morphology. The microsphere is a remarkably stable structure in that it can be centrifuged at high speed, lyophilized, heated etc. without destroying its integrity. On the other hand, shaking a suspension of microspheres results in fragmentation and increase in number of spheres, which under certain circumstances may be mistaken as division.

Even the contemporaries of Fox vividly criticized in the 1970s, his euphoric view of proteinoid microspheres as precursor of living cells endowed with the "talent" of proliferation. In the twenty-first century, the proteinoid microspheres vanished completely from the discussion of CE theories, in as much as neither Fox nor anybody else had shown that specific chemical reactions can occur inside.

A more sophisticated continuation of Oparin's first compartment theory was elaborated by the biochemist T. Ganti in the years after 1971 (Ganti 1971, 1974, 1979, 2003). He designed and advocated a s. c. "Chemoton Theory," which was focused on the evolution of a specific metabolism inside vesicles. The following text is reproduced from his book *Principles of life* (2003, pp. xii and xiii):

> [...] all living must contain an autocatalytic network comprised of reversible reactions and similarly autocatalytic, but unidirectional reaction system of irreversible reactions. The specific coupling of the two creates a functional system or "machine" which unlike machines created by humans manipulates the energy that flows through it in chemical rather than mechanical or electric ways.

And:

> In this book the term "chemotons" was coined for the simplest chemical "machine" which shows the generally established characteristic of live.

With this statement, his concept may be understood as the beginning of the "minimal cell theory" (Luisi 2006) and thus his "Chemoton" theory does rather belong to the "top-down" approach to CE than to the bottom-up counterpart.

Another early compartment concept was published by the biologist C. Woese in 1979. He hypothesized that droplets in the atmosphere played the role of compartments and he rejected the Oparin/Haldane hypothesis that a primordial soup covering a significant part of the Erath's surface was the birthplace of CE:

> The widely accepted Oparin thesis that the origin and early evolution of live seems sufficiently far from the true state as to be considered incorrect. It is proposed that life on Earth actually arose in the planet's atmosphere, however an atmosphere very different from the present one. Because an extremely warm surface, the early Earth may have possessed no liquid surface water, its water being partitioned between a molten crust and a fairly dense atmosphere. Early preliving systems are taken to arise in the droplet phase in such an atmosphere. The early Earth which resembles Venus then and to some extent now, underwent a transition to its present conditions largely as a result of an evolution of a methanogenic metabolism.

Woese did not disclose concrete comments on the chemistry that might have taken place inside the droplets, and he did not contribute any experiments.

A further theoretical elaboration of Oparin's coacervate theory was contributed by F. Dyson in 1983. He developed a mathematical model of a compartment specific metabolism based on the following assumptions:

1. Cells came first, enzymes later and genes much later.
2. A cell is a compartment containing a population of various biopolymers which are different from those of the environment.

He explained then his approach as follows:

> This chapter describes my own attempt to understand the Oparin theory of the origin of life. The general theory begins with an abstract multidimensional space of molecular populations [::::]: The decisive events in a theory of the origin of metabolism are the rare statistical jumps, when a population in one quasi stationary state happens to undergo a succession of chemical reactions which push it up against the gradient of probability over a barrier and down into another quasi stationary state. If an initial state is disorganized and the final state is organized, the jump may be considered to be a model for the origin of metabolism. In a complete theory about the origin of life, it is likely that there would be several such jumps, each jump taking a population of molecules to a new quasi-stationary state having a more complete structure than the previous state.

This approach looks interesting and promising, but in CE research like in any other kind of chemical research, nice mathematical models cannot substitute experiments. A weak point of the Dyson–Oparin theory is, for example, the assumption that the presence of homochiral amino oligopeptides, proteins, and saccharides were available.

Concrete ideas about the chemical structure and biological function of compartments were described by H. J. Morowitz (1992). He postulated an early formation of vesicles built up by membranes consisting of layers of lipids:

> Many early theories of biogenesis have a postulated elaborate evolutionary development in a unpartitioned, aqueous environment – the "primordial soup" or the Oparin ocean. This places an extraordinary burden to counter diffusion and the other dissipative consequences of the Secondary Law of Thermodynamics. At what stage of biogenesis partition occurred is an open question, but one answer of early rather than late provides some obvious advantages. One of the primary features that characterizes the approach of the monograph is its strong focus on the notion that membranes and vesicle formation constituted one of the early stages and involve an abrupt transition from the homogeneous to the heterogeneous domain [....]. I shall argue that that transition to vesicles led to the directed chemistry necessary for biogenesis [....]. It is for this reason that the title of the book is "The beginning of Life". It is the closure of an amphiphilic bilayer membrane into a vesicle that prevents a discrete transition from nonlife to life.

H. J. Morowitz outed himself as a fan of a directed evolution (s. Sect. 1.1), but he expected all stages of the CE can be proven by experiments, because the course of the chemical evolution is predetermined in the structure of the orbitals of atoms and molecules. However, the numerous model experiments reviewed in Chaps. 4, 5, 6, 7, 8, 9, 10, and 11 did not only demonstrate the absence of the postulated evidence, on the contrary, they provided results that are in contradiction to the concept of a directed, deterministic evolution.

The idea that an early formation of micelles and vesicles formed by monolayers or bilayers of lipid membranes (s. Fig. 2.1) was decisive for course, and success of the CE also stimulated numerous experimental studies. A major contribution was elaborated by D. W. Deamer and coworkers beginning in 1977 with the work of W. Hargreaves (Hargreaves et al. 1977, Hargreaves and Deamer 1978. The first step in this direction concerned esterification of glycerol or glycerol mono-ethers with dodecanoic acid catalyzed by kaolin or silicic acid by means of cyanamide and sodium hydrogen phosphate as condensing agents (Hargreaves et al. 1977). Those authors had positive results, but quantification of the reactions was not feasible at that time. In subsequent experiments, esterification of glycerol with dodecanoic acid in the absence of catalysts was studied. The yields of mono-esters varied with the reaction conditions (Apel and Deamer 2005). In further model experiments, aqueous solutions of pyruvic acid were heated up to temperatures around 250 °C under a pressure of 200 GPa. These experiments had the purpose to simulate the conditions of hydrothermal fountains (e.g., "black smokers"). Hazen and Deamer

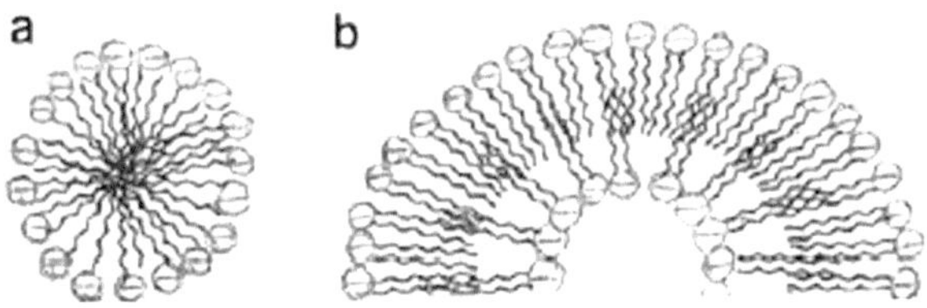

Fig. 2.1 Anionic amphiphiles (e.g. anions of alkanoic acids) which form (**a**) a simple micelle in water or (**b**) a dounle layer of a vesicle

(2007) found vesicle forming components in the complex reaction mixture and speculated about formation of alkanoic acids similar to those extracted from the Murchison meteorite.

Deamer and coworkers also reported that the addition of non-ionic amphiphiles stabilized the double layer of aqueous solution of decanoic acid. Their sensitivity against variation of the pH was reduced, and addition of metal ions, such as Mg^{2+}, Ca^{2+}, and Fe^{2+} did not cause coagulation (Apel and Deamer 2005). An even better stabilization of the alkanoic acid double layers was found by Namani and Deamer (2008), when decylamine was combined with decanoic acid. Surprisingly a stabilizing effect was also observed, when nucleobases or monosaccharides were added to bilayers of decanoic acid. The addition of these bio-monomers prevented flocculation of the decanoic acid when divalent metal ions were added. Interestingly, D-ribose showed a stronger stabilizing effect than other mono-saccharides (Black et al. 2013). Further studies concerned the influence of functional aromatic compounds on bilayers of decanoic acid. 1-Hydroxypyrene, anthracene 9-carboxylic acid, and 1,4-chrysenequinone were studied, because these aromatics were found in meteorites (although in traces only). Groen et al. (2012) concluded:

> Vesicle size distribution and vesicle concentration are largely unaffected by PAH incorporation, but 1-hydroxypyrene, 9-carboxyanthracene lowered the permeability of fatty acid bilayers to small solutes up to 4-fold. These data present the first indication of a cholesteric-like stabilization effect of oxidized PAH derivatives in a simulated prebiotic membrane.

Finally, the Deamer group studied the influence of phospholipid layers, nanofilms, crystalline ammonium chloride, and montmorillonite on the base pairing of AMP and UMP mixtures (Himbert et al. 2016). In that study, the phospholipid layers showed the weakest effect. The studies of amphiphilic substances extracted from meteorites conducted by the Deamer group will be discussed in Chap. 9 (Deamer 1985, 2005). Further studies concerning the stability of vesicles in the presence of inorgsnin ions or organic amphiphiles were also contribuzed by other research groups (Monnard et al. 2002, Fujikawa et al. 2005).

As early as in 1987, Rao et al. had also pointed out that membranes based on amphiphilic double layers might have played a major role for the formation of the first cells (Rao et al. 1987). Their studies of cyanamide promoted condensation reactions of phosphatidic acid with ethanol amine had a moderate success. However, none of the compartment theories discussed before have made any contribution to the answer, how proteins or RNA evolved in the course of the CE.

However, experimental progress in this direction was contributed by several research groups after 1980. In 1981, Kuneida et al. reported that prAAs esterified with lipophilic alcohols (a in Scheme 2.6) condense easily yielding oligopeptides, when the reaction condition favors formation of micelles (Kuneida et al. 1981). In contrast, condensation of prAA alkyl esters in homogeneous solution yields cyclic dimers (2,5-diketopiperazines) as main reaction products. The high concentration of reactive end groups in lipid double layers also favored polycondensation of the tripeptides formulated under b in Scheme 2.6, because their

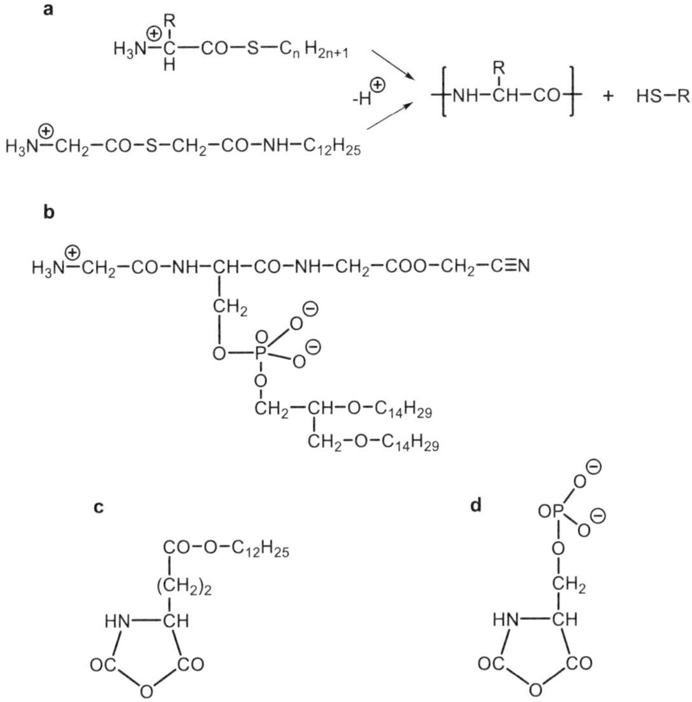

Scheme 2.6 Monomers that have been successfully polycondensed or polymerized within cells

lipophilic side chains enabled their embedding in membranes (Nishikawa et al. 1993). Blocher et al. (1999) compared polycondensations of several dipeptides promoted by EEDQ (2-ethoxy-1-ethoxycarbonyl dihydroxyquinoline) in the presence of liposome membranes. The highest yields, up to 70%, were obtained with H-Trp-Trp-OH, whereas the less lipophilic dipeptides H-Trp-Gly-OH, H-Trp-Glu-OH and H-Trp-Asp-OH preferentially yielded cyclic dimers. Apparently, the most lipophilic dipeptide was enriched in the membrane which favored condensation of neighboring dimers and hindered cyclization.

Three research groups worked on ROP of NCAs (s. Sect. 3.3) derived from prAAs inside and outside vesicles. At first, Shibata et al. (1986) described the ROP of γ-dodecyl-L-Glu-NCA (c in Scheme 2.6) and the ROP of γ-benzyl-D-Glu-NCA in the presence of liposomes and compared these polymerizations with ROPs in dioxane, a comparison which did not make sense. Only DPs around 9 were found in the water/liposome system and, as predictable, higher DPs in dioxane. Those authors should have compared ROPs in water in the presence and absence of liposomes. More successful were the experiments of Böhler et al. (1996). Three different NCAs having negatively charged side chains, namely the NCAs of Asp, Glu, and phosphoSer (**d** in Scheme 2.6) were transformed into NCAs by means of carbonyldiimidazole (CDI) and polymerized in situ. In the presence of cationic liposomes, such as cetyl trimethyl ammonium chloride, high yields and high DPs were obtained, and those authors assumed that enrichment of the NCAs in the lipophilic membrane was responsible for their success.

The group of P. L. Luisi (Blocher et al. 1999, 2000; Hilz et al. 2001) studied the ROPs of Trp-NCA or Trp + CDI in the presence or absence of liposomes. Since Trp has a lipophilic side chain, the presence of liposomes enhanced both yields and DPs. When D,L-Trp was used no preference of the D- or L-enantiomer was observed, but homochiral sequences were preferentially formed (s. Sect. 10.3). In summary, all these experiments demonstrated for the first time that lipophilic membranes may have a positive influence on the formation of certain oligo- or polypepetides. Unfortunately, copolypeptides having high molecular weights that might be considered as models of proteins could not be achieved by polycondensation and copolymerizations of NCAs cannot have played an important role in CE for reasons discussed in Sect. 3.3. Concerning the origin of the phospholipids, Luisi said in his book *The emergence of Life* (2006, p.206):

> Phospholipids are the main constituents of most biological membranes and are produced in modern cells by complex enzymatic processes: as such they cannot be considered prebiotic compounds.

In the years after 2002, the research group of J. W. Sczostak has launched an extensive research program concerning formation and function of lipid membranes under the aspect of their role in CE. The numerous results were published as follows: Hanczyc and Szostak (2004), Hanczyc et al. (2003, 2007), Chen et al. (2005, 2009), Chen and Szostak (2009a, b), Mansy et al. (2008), Many and Szostak (2008), Zhu and Szostak (2009a, b), Budin and Szostak (2011), Budin et al. (2009, 2012, 2014), Abdemale and Szostak (2013a, b), Abdemale et al. (2016), Kanat et al. (2015). In this connection the question was studied, if and under which conditions vesicles may undergo proliferation. Perhaps the most interesting result was the finding that a photochemical initiated redox process stimulated the proliferation of vesicles (Zhu et al. 2012). A further focus of the Szostak group were chemical reactions inside the vesicles, for example, RNA-catalyzed syntheses. (Chen et al. 2005). Furthermore, template-supported non-enzymatic syntheses of RNAs by polycondensation of nucleotide imidazolides as monomers (Mansy et al. 2008; Abdemale and Szostak 2013a, b), (a detailed description of this synthetic method is given in Sect. 7.3). Enzyme-catalyzed reactions in vesicles were also reported by Ishikawa et al. (2004).

In summary, the afore-mentioned studies of syntheses and properties of micelles and vesicles suggest that at least in a later stage of CE, these organic compartments played a decisive role for the emergence of the first living cell. However, this conclusion does not exclude that inorganic compartments played an important role at beginning of the CE favoring syntheses of bio-monomers from simple starting materials such as CO_2, CO, CH_4 N_2, NH_3, and H_2O. A "transfer of information" and catalytic properties from inorganic to organic compartments may also be taken into account. The "Genetic takeover" postulated by Cairns-Smith for the "Silicate World" is an extreme speculation (s. next section), but an interesting experimental result in this direction was published by Hanczyc et al. (2007). Those authors demonstrated that certain mineral surfaces can favor formation of lipid membranes and vesicles from fatty acids. However, the question, how unsubstituted fatty acids or linear alkanols were formed under prebiotic conditions remained an unanswered problem. A discussion of this problem is presented in Chap. 9 in connection with the role of chemical reaction in space. Finally, it should be mentioned that all the hypotheses and experiments presented in the following sections belong more or less to the compartment theories.

2.5 The Clay World

The "Clay World" hypothesis was elaborated by the theoretical biologist AS. G. Cairns-Smith in 1970s and summarized in a book entitled *The Genetic Takeover* published in 1982 (Cairns-Smith 1982). In agreement with the label, "theoretical" biologist Cairns-Smith never contributed experiments to his vision. However, his descriptions and comments on structure and properties of clay are based on experimental results of other research groups, and from this point of view, his concept is a hypothesis and not mere speculation. In fact, his concept is the most visionary and original contribution to the field of hypotheses about the emergence of life via chemical evolution.

Silicates cover more than 95% of the dry surface of the Earth, and this scenario has certainly not changed since the Earth formed a solid crust upon cooling down during the first stage of its formation. The architecture of silicates may vary over a broad range from linear single chains over band-like structures formed by two parallel chains to sheet-like structures and finally to three-dimensional networks. The three-dimensional arrangement of sheet-like silicates is variable, because the distance between neighboring sheets may vary with the consequence that a pile of sheets may host inorganic or organic molecules between the sheets. Networks may form pores and compartments of different size and three-dimensional form, which according to Cairns-Smith can play the role of mini-reactors. The chemical structure is further complicated by incorporation of aluminate groups. When Al replaces Si in a tetrahedra formed by four oxygen atoms, the deficit of one electron relative to Si has the consequence that the AlO4 group must have a negative charge. In other words, aluminosilicate chains are negatively charged polyelectrolytes which may be neutralized by various cations. Na^+, K^+, Mg^{2+}, and Ca^{2+} are certainly the most abundant soluble cations in the ocean and on the surface of the Earth, but other cations such as Fe^{2+}, Co^{2+}, Ni^{2+}, Cu^{2+}, or Zn^{2+} may also be present in small quantities. All these elements are present on or near the surface of the Earth in the form of sulfide minerals or as oxides after oxidation of the sulfides. With exception of Ni^{2+}, the other cations play an important role in biocatalysts of living organisms, and thus, it is not far-fetched, when Cairns-Smith assumes that alumosilicate layers or compartments containing these cations might have played the role of chemical reactors. Depending on the spatial situation and nature of the cations present, such mini-reactors might have developed a high selectivity for individual reaction pathways.

Not only Cairns-Smith, but also the crystallographer J. D. Bernal (1967) postulated that silicates played a decisive role for the success of chemical evolution. He interpreted their role primarily as "harbors" or pools that enabled extraction of organic compound from the primordial soup, and enabled their reactions by enhancement of their concentration.

The concept of Cairns-Smith describes the progress of the evolution in four steps:

1. Emergence of the inorganic clay chemistry.
2. Interaction with organic molecules and catalysis of syntheses favoring biomonomers.
3. Transfer of the inorganic (mineral) reproduction mechanism on organic oligomers and polymers such as RNA and proteins.
4. Substitution of the "mineral life" by the more efficient biopolymers and emergence of the first cell.

Cairns-Smith was aware that his vision was a provocation for the majority of scientist, and in the introduction of his book *the genetic Takeover*, he wrote:

> The main idea in this book is that the first organisms on Earth had an altogether different chemistry from ours, that they had a solid state biochemistry. This idea was already in the opening sentence of the abstract of my first paper on the origin of life (1966). It is proposed that life on Earth evolved though natural selection from inorganic crystals. It sounds an odd idea, but I meant it literally. I still do..[…]. I have persuade you that the very first organism need not have contained organic materials..[..]. I have to provide an account, at least in principle, and in as much detail as possible, as to how early evolutionary processes should have transformed organisms with one kind of central control machinery into organisms based on control structures of an altogether different kind (This is genetic takeover). I should explain too, why evolution should have had such oblivious beginning.

Cairns-Smith assumed that clay forms micro- and nano-particles by erosion of mountains. They were transported by rainfalls, creeks, and rivers, thereby their shape changed more or less and their chemical structure was slightly modified. Silicate crystals, which have a lamellar structure, grow by lateral addition of dissolved silicate building blocks. With increasing size, the probability increases that small pieces of the lamella break off and serve as nucleus for the growth of a new large lamella. In other words, clay was capable of proliferation.

A first, primitive inheritance mechanism resulted from the transfer of specific properties (e.g., defects) of the "mother lamella" to the broken off lamella, which, in turn, maintained this property in the course of its growing process. In this way, a next generation of identical lamellae having identical properties different from those of other lamellae might have been formed. For example, one might consider that in different areas of a large lamella, the sequence of Al and Si atoms was somewhat different. When this lamella broke into two or three pieces, these pieces had different sequences of Al and Si atoms, and their proliferation produced three populations of lamella which differed by their ratios and sequences of Al and Si atoms.

The next step in the evolution of the clay world was formation of three-dimensional architectures containing compartments in the form of card houses as

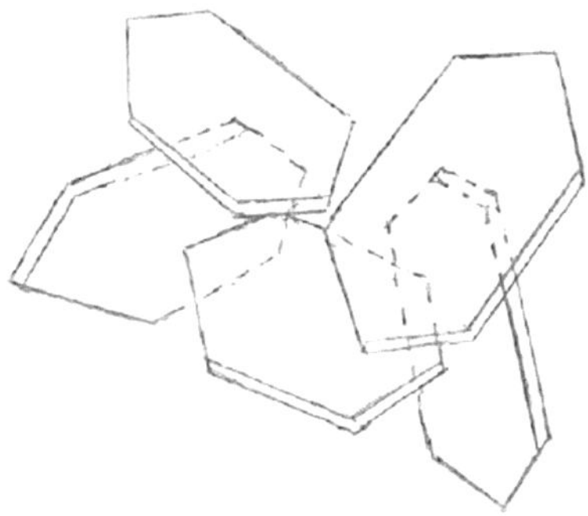

Fig. 2.2 Model of a cart-house, which may be formed according to Cairns-Smith (1982) via electronic interaction of positively charged edges of silicate crystallites with their negatively charged surfaces

illustrated by the simplified Scheme of Fig. 2.2 (Fig. 2.3 in Cairns-Smith's book). These compartments were thought to become the role of reactors, which differed by their specificity for chemical reactions, depending on their size, their shape, and their content of cations. Cairns-Smith speculated about syntheses of prAAs and nucleobases starting out from the metabolism of living organisms. This concept requires that different reactors were in contact with each other and exchanged their reaction products, so that the final goal, the syntheses of biomonomers and biopolymers required for the formation of the first cell, was finally achieved. Each reactor allowed for the influx of the necessary starting materials and for the disappearance of useless by-products. In other words, the clay developed an intelligent production center, where all parts cooperated in such a way that the final goal, the first cell was formed.

Latest at this point, massive critique is necessary. First, there are no model reactions which support the idea that clay was capable of developing intelligent production lines for complex organic compounds, above all biomonomers and biopolymers. It is indeed known that silicates with layer structure, e.g., Biotite and Montmorillonite, favor formation of longer oligopeptide chains than polycondensations of the same monomers in solution (s. Chap. 6). However, these results do not suffice to justify the assumption that clay was capable to organize reactors having different specificities for syntheses of the various biomonomers needed for the production of proteins or RNA and DNA, in as much as evidence for the individual types of reactors based on model reactions is also missing. Second, there is no justification why a directed production process with intelligent coordination of several different reactors emerges from an undirected chemical evolution. The third argument against the "Clay World," the total lack of any evidence from geological research is of minor importance, because clay sediments having an age of >3.5 billion years have not been found yet. In summary, the vision of the self-organization of a "clay industry" specialized on the production of all the components needed for successful chemical evolution is a fantastic day dream.

The "Clay World" hypothesis still includes two more steps in direction of the first cell. Production of biomonomers and polymers alone does not suffice. The vision of Cairns-Smith includes the assumption that the polypetides and RNAs produced in the clay factories had excellent catalytic properties and the capability of information storage. Furthermore, they were capable and willing to cooperate with the goal to produce the first cell, and on this way, they have finally killed their parents, the clay factories. This scenario requires another massive critique, which concerns the core of the Cairns-Smith hypothesis. The assumption of a "Genetic Takeover" is exclusively based on the knowledge that the biochemistry of all living and fossil organisms extant today involves organic compounds. As demonstrated also in other Chapters of this book, such "*a posteriori* arguments" can be used as justification for various strange hypotheses. A "genetic takeover" is absolutely unlikely, when an undirected evolution is the basic concept. Why should the clay organisms have created a totally different type of biochemistry and a totally different class of organisms, which, in turn, destroyed their origin, or in other words, which killed their parents. This question is connected to the following question. Even when organisms working with an organic biochemistry were created, why should these organic organisms have destroyed the "Clay World." There was no competition for nutrition and almost no competition for space. The "clay organisms" needed soluble silicates for their nutrition, whereas the organic organisms needed organic compounds of the primordial soup as feedstock. Without competition, no selection can occur resulting in the disappearance of the inferior sort of organisms.

Furthermore, the influence of catastrophes needs to be taken-into-account, and catastrophes occurred certainly more frequently on the early Earth than today. Regardless, if a rapid raise or reduction of the temperature is considered, a change of the atmospheric pressure, a change of the chemical composition of the atmosphere or impacts of asteroids, the "clay organisms" were more robust than the organic organisms, and thus, had a greater chance to survive and to improve their system. Any catastrophe driven selection mechanism favored the "Clay World." Hence, there is neither a convincing theoretical argument nor any experimental argument supporting the hypothesis of a genetic takeover of the "Clay World." In a philosophical comment, R. Shapiro formulated the following conclusion:

> We must wait for any final answer for clay life. If in fact we began in this way, it would be one of the most satisfying answers. We inhabit this planet and we use its resources. Our bodies are placed into the earth when we pass on. How fitting, when we were born of this soil as well, as suggested in Genesis 2.6-2.7. But there went up a mist from the Earth and watered the whole face of the ground. And the Lord God formed the man out of the dust of the ground and breathed into his nostrils the breath of life, and man became a living soul.

2.6 The Thioester World, the Iron-Sulfide World and the "Black-Smokers"

Three authors proposed the hypothesis that sulfur salts and sulfur compound played a decisive role in the early stage of the chemical evolution. The term "Thioester World" was coined by the Nobel Prize Laureate of medicine (1974) C. DeDuve (1991, 1998). He assumed that thioesters of carboxylic acids, such as acetic acid or prAAs played key role as energy-rich intermediates that promoted most condensation reactions. His hypothesis was stimulated on the one hand, by the finding that primitive organisms having a metabolism based on a sulfur/sulfide redox reaction live in the neighborhood of the hot hydrothermal vents called "black smokers." On the other hand, his view was stimulated by the finding that acetyl-coenzyme which belongs to the human metabolism contains acetyl thioester group as active species. However, DeDuve was physiologist and did not perform chemical experiments. Yet chemical experiments dealing with the synthesis of carboxylic acid thioesters (including thioesters of prAAs) under plausible prebiotic conditions were conducted at the same time by A. Weber (1981, 1984, 1998). He prepared thioesters from mercaptans and aldehydes or α-keto acids via redox reactions with Fe^{3+} ions as oxidizing agents. Yet, experiments yielding oligopeptides or oligonucleotides were not performed. Therefore, the relevance of these experiments for the course of the CE remained obscure.

Iron sulfide plays a decisive role in the hypothesis promoted by the chemist G. Wächtershäuser (1988). This "Ion sulfide World" hypothesis and the "Thioester Worlds" have in common that they are connected to existence of "black smokers" which exhale FeS particles, but in the "Iron sulfide World" concept, the iron sulfide plays the role of catalyst and not the role of reagent. When the "Iron sulfide World" is compared with the "Clay World" difference consist in the temperature and mainly in the nature of the "working place" of the molecular evolution. Whereas clay hosts almost all chemical reactions in pores and compartments, most chemical reactions of the "Iron sulfide World" proceed on the surface of pyrite (FeS_2) crystals. Since pyrite and its alternative crystal modification, marcasite, belong to the most abundant heavy metal ores on the surface of the Earth, this choice of the evolutionary working place cannot a priori be criticized.

Both the clay world and the "Iron sulfide World" have one characteristic in common. Both hypotheses focus primarily on the evolution of chemical reactions in general and synthetic pathways for biomolecules, in particular, and do not favor the evolution of biopolymers capable of self-replication and evolution of a genetic system. Furthermore, their authors understood their hypotheses not as contribution to the Oparin or Haldane hypotheses of evolution originating from a primordial soup, but as alternatives. This view was stimulated by increasing critique of the Oparin/Haldane soup hypotheses which came up after 1980. The original comment of Wächtershäuser (2003) reads:

> It is occasionally suggested that experiments within the Iron sulfide World demonstrate merely yet another source of organics for the prebiotic soup. This is a misconception. The new finding [formation of pyruvic acid at high temperature and high pressure] drives this point home. Pyruvic acid is too unstable to ever be considered as a slowly accumulating component in a prebiotic broth. The prebiotic broth theory and the Iron sulfide World theory are incompatible. The prebiotic broth experiments are parallel experiments that are producing a greater medley of potential broth ingredients. Therefore, the maxim of the prebiotic broth theory is "order out of chaos". In contrast, the Iron Sulfide World experiments are serial, aimed at long reaction cascades and catalytic feed-back (metabolism) from the start. The maxim of the Iron Sulfide World theory should, therefore, be "order out of order out of order."

Wächtershäuser describes his concept as emergence and evolution of two-dimensional autotrophic reaction cycles. The surfaces of the pyrite crystals which host the chemical reactions affect the two-dimentionality of the entire iron sulfide world. Reactions on the surface ofFeS_2 crystals have the advantage that the reaction partners have much higher concentration than in a prebiotic soup. Furthermore, the FeS_2 crystal has catalytic properties, which vary with the presence of "impurities," such as Ni^{2+} or Cu^{2+} ions. In combination with H_2S which evaporates from volcanoes (including hot vents in the oceans), several redox reactions become possible which generate energy. Sunlight, radioactivity, or electric discharges are not required as external sources of energy. For this reason, the iron sulfide world chemistry may be called autotrophic, in contrast to almost all model experiments concerning formation of biomonomers in the primordial soup (s. Chaps. 4 and 5). Hence, the iron sulfide world chemistry may occur in the dark, for example, on the underground of oceans. Warm or even hot areas providing H_2S, such as the "black smokers" are particularly favorable.

It is an important advantage of the iron–sulfide world hypothesis relative to the clay world that several postulated reaction steps are supported by model experiments. Table 2.1 summarizes the model experiments elaborated up to the year 2003. However, at this point, the crucial question arises, was it possible and how was it possible that a two-dimensional metabolism limited to small number of specific reactions limited to sulfur chemistry was transformed into a three-dimensional chemistry focused on the evolution of proteins, nucleic acids, and vesicles containing a metabolism different from that of the surrounding medium. For this dramatic change, experimental evidence does not exist, only speculations of Wächtershäuser may be cited (Wächtershäuser 1988):

> The genetic machinery of the cells develops from the surface metabolism precursors with catalytic imidazole residues glycosidically bonded to a poly(hemiacetal) backbone of (surface adhering) phosphotrioses. It produces self-folding enzymes which compete with the mineral surface for bonding the metabolic constituents. In this stage evolution becomes double tracked, an evolution of metabolic pathways and one of the bonding surfaces for their constituents. In the third stage the pyrite support is abandoned and true cellular organisms arise which become free to conquer 3-dimensonal space.

Table 2.1 Reactions of the iron sulfide world "metabolism" supported by model experiments

Reaction	Catalyst	Temp. (°C)	Reference
$CO > HCO_2H$	(Fe, Ni)S	100	Huber and Wächtershäuser (1997)
$CO > CH_3SH$	(Fe, Ni)S	100	Huber and Wächtershäuser (1997)
$CO > CH_3CO_2H$	(Fe, Ni)S	100	Huber and Wächtershäuser (1997)
$HCO_2H > CH_3COCO_2H$	FeS	250	Cody et al. (2000)
$CH_3SH >$ CH3CO-SCH_3	(Fe, Ni)S	100	Huber and Wächtershäuser (1997)
$CH_3COCO_2H > NH_2CHRCO_2H$	FeS	100	Huber and Wächtershäuser (1997)
$NH_2CHRCO_2H >$ Oligopeptides	(Fe, Ni)s	100	Huber and Wächtershäuser (1998)

And in 1994 (Wächtershäuser 1994):

> Now it is important to realize that a synthetic autocatalytic cycle has an inherent tendency towards higher complexity. That means an inherent tendency towards an increasing number of reaction possibilities. Thus, life creates its own prospects. This is the physical basis of the interpretation of evolution as a process of self-liberations.

These nice visions provoke the following comment. One may accept that with time number and diversity of the chemical reactions and reaction products have increased. However, this increase of quantity does not necessarily include a significant increase of complexity, such as formation of proteins and RNAs. Furthermore, even when hundreds, thousands, and hundred thousands of copolypeptides or copolynucleotides are formed, these copolymers do not necessarily have the wish to cooperate and to create a living cell. At this point, Wächtershäuser's vision drifts away from a serious scientific hypothesis toward wishful thinking. However, at first glance, a surprising experimental result seemed to support the iron–sulfide world hypothesis. Huber and Wächtershäuser discovered in 1998 that dipetides of prAAs may be formed (in tiny quantities), when CO and H_2S (or CH_3SH) react with each other at 100 °C in the presence of FeS_2 doped with low concentrations of N^{2+} ions (Huber and Wächtershäuser 1998). The authors worked with aqueous solutions at pH values in the range of 7.5–10.1. Under these harsh conditions, it is not surprising that oligopeptides did not accumulate, because of rapid hydrolysis.

The measured quantities of dipeptides were interpreted by Huber and Wächtershäuser as components of a dynamic chemical equilibrium, which was seen as analogy to metabolic cycles of living organisms (Huber et al. 2003). This interpretation may satisfy a fan of metabolic evolution, but these results demonstrate primarily that under the chosen reaction conditions, higher oligopeptides or polypeptides cannot be formed. A second and likewise important point of critique concerns the problem of racemization. Monomeric prAAs-like prAA units in peptides suffer from rapid racemization, when confronted with high temperatures and alkaline pH values. Even in water with neutral pH, several prAAs racemize within a few weeks, when exposed to a temperature around 100 °C (s. Chap. 10). A period of a few weeks is a blink of an eye compared to the many millions of years required for a successful chemical evolution. In other words, a chemical evolution toward homochiral biopolymers was impossible under these conditions. His kind of critique concerns any hypothesis of chemical evolution which is based on an environment with high temperatures (>70 °C) and/or strongly acidic or alkaline (pH >7.5) water. Model experiments, which demonstrate, how homochiral polypeptides evolve from mixtures of racemic prAAs would disprove this critique, but such model experiments don't exist.

The author of this book is by no means the only criticist of the iron sulfide world as basis of a complete successful chemical evolution toward a living cell. As early as 1981, De Duve and Miller have already criticized the early version of Wächtershäuser's concept. Wächtershäuser has carefully analyzed their critique (Wächtershäuser 1994) and demonstrated that a significant part of the critical arguments was not justified. However, L. E. Orgel (2000, 2003) published later further critical arguments, which were not disproved by Wächtershäuser, and the problem of racemization formulated by the author above, was not even included in the critique of Orgel. All facts and arguments together let the author conclude that the "Iron Sulfide World" is a highly original and fascinating concept, which supported by several model experiments can explain how cascades of certain chemical reactions yielding organic compounds may evolve. But this scenario does not suffice to establish convincing hypothesis of a chemical evolution resulting in the formation of the first living cell.

2.7 The "White Smokers"

The term "white smokers" is here used as a nickname for hydrothermal fountains of hot sea water that do not expel a dark smoke like the "black smokers" (which liberate FeS and NiS in the form of fine powders), but eject water rich in various minerals at temperatures in the range of 50–80 °C. These white smokers are typically located in the close neighborhood of stress and subduction zones of the Earth crust, and thus, they may also be located in the neighborhood of "black smokers." However, their properties differ largely from those of the "black smokers" in a way which stimulates biologists to consider "white smokers" as potential birthplace of CE. The life time of "black smokers" amounts to only several decades, because the hot spots which heat them up are in permanent motion and migrate away. In contrast, the "white smokers" may be alive for several thousands of years. Furthermore, the flow of water is relatively slow when compared with the Jetstream-like ejection of water from the "black smokers," which occurs under high pressure. Hence, an accumulation of organic molecules, biopolymers, and primitive cells in the pores of the walls of the "white smokers" is much more likely than an accumulation in the walls of the "black smokers." However, another difference is unfavorable for a homochiral CE, namely the pH, which is slightly acidic in the case of "black smokers, but strongly alkaline (pH 9-11) in the case of white smokers" (s. Sect. 10.3).

Over the past 20 years, numerous biologists and geologists have expressed their sympathy for the "white smokers" as the cradle of CE and birthplace of the first cells (Früh-Green et al. 2004; Holm et al. 2006; Schulte et al. 2006; Martin and Russel 2007; Martin et al. 2008; Russell and Arndt 2005; Russell et al. 1994, 2010; Mulkidjanian 2009; Sleep et al. 2011; Sousa et al. 2013). In connection with the properties of "white smokers," the biologist N. Lane has published the most comprehensive and complete type of compartment theory in a book entitled *The vital Question* (Lane 2015). In this book, he presented the following arguments in favor of his hypothesis.

First, the walls of the white smokers consist of highly porous serpentinite. Serpentinite is a metamorphous kind of stone which is formed by the impact of hot water on island silicates such as olivine or band silicates, such as pyroxene.

Second, the walls of the pores may contain calcium carbonate and/or traces of heavy metal salts, such as FeS, NiS, CuS, or ZnS. These heavy metal ions might have played a decisive role as catalysts, above all, for redox reactions, in as much as Fe and cu ions still play today an important role as components of enzymes active in the respiratory chain of all higher organisms.

Third, in 2010, Russel had discussed that formation of serpentinite in hydrothermal fountains is accompanied by liberation of H_2, so that by reaction with N_2 and CO_2 (catalyzed by FeS and NiS), ammonia, methane, methanol, formic acid, and further more complex organic compounds might be formed (Russell et al. 2010).

Fourth, part of the pores is protected against a rapid exchange of their content with seawater by a network of thin silicate fibers which play the role of a membrane. Hence, an accumulation of larger organic compounds may occur (experimentally proven), which, in turn, enable a variety of chemical reactions different from those in the surrounding seawater. This argument reminds the reader of the silicate world of Cairns-Smith and the decisive role of FeS+NiS catalysts is borrowed from Wächtershäuser's iron-sulfide world.

Fifth, the silicate membranes are assumed to be capable to separate to some extent the alkaline water of the "white smoker" from the surrounding less alkaline seawater. This proton gradient might have stimulated the formation of an adenosine triphosphate (ATP) "production plant." ATP is used in all cells as the basic fuel for almost any kind of chemical reactions. In the mitochondria of living organisms, the activity of the ATP synthesizing unit is also fueled by a proton gradient, but this gradient needs to be maintained in the respiratory chain by an energy consuming proton pumping system consisting of proteins. Now Lane hypothesizes that redox reaction between CO_2 and H_2 in the pores of the serpentinite have provided the energy needed for formation and maintenance of the first cells. As the first step of this chain of redox reactions formation of acetic acid was taken into account. In contrast to the later bacteria, the archaea leaned to continue the redox processes up to the production of methane. Furthermore, Lane assumed that acetyl thioesters (CH_3CO-SR) and acetyl phosphate (CH_3CO-O-$PO3^{2-)}$ played the role of energy rich fuel before ATP took over this role.

The above list of arguments is only a short version of much more detailed description of Lane's concept which is focused on the central role of formation and transport of chemical energy as basis and limitation of the chemical and biological evolution. So Lane, Martin, and Russel have designed a plausible scenario, how and why the first cells were formed and existed in the pores of serpentinite. Furthermore, they designed a plausible scenario for the evolution of archaea and bacteria from the first primitive cells. However, what is largely missing is a comment how the chemical basis of their scenario was formed evolved out of a few simple starting materials such as NH_3, CO_2, CH_4, and N_2. The most important problems are summarized and discussed below.

1. Several prAAs, such as lysine, arginine, histidine, tryptophan, and cysteine, were not pound in model experiments (s. Chap. 4) nor was it possible to extract them from carbonaceous meteorites (Sect. 9.4). However, these prAAs were needed for the synthesis of the ATP "production plant," but where is they come from?

2. How and when were the prAAs selected from a pool of 80–90 amino and imino acids that have possible existed in the primordial soup as evidenced by model experiments and extraction of meteorites?
3. How and why were the nucleobases typical for RNA selected from a pool of ten or more nitrogen heterocycles that may have existed in the primordial soup? A couple of them are capable of base pairing (xanthine, hypoxanthine, diaminopurine, methyl cytosine, dimethyl uracil).
4. How were small reactive molecules (e.g., aldehydes, NH_3, or formic acid) that might have prevented formation of proteins and RNAs kept outside the pores producing the ATP synthetase? Furthermore, the assumed ATP precursors acetyl thioesters and acetyl phosphate react rapidly with amino groups and thus, stop any kind of peptide or protein synthesis.
5. In the case of asparagine and glutamine, the CO–NH_2 groups are rapidly hydrolyzed at pH >9 in hot water. Hence, they were not available for syntheses of proteins inside the "white smokes" even when small quantities existed in the surrounding prebiotic soup.
6. A particular important problem was the high pH in combination with temperatures above 50 °C. Under such harsh conditions, several prAAs, such as Asp, Glu, Ph, and His racemize within a few hours and the other prAAs racemize within a few days (s. Sect. 10.3). In other words, an evolution of homochiral peptides and proteins was impossible under the conditions offered by the "white smokers."

It is understandable that biologists are not so familiar with all the chemical problems faced by a theory of chemical evolution as a chemist should be. However, his point means that more cooperation between chemists and biologist or geologists is needed to achieve a convincing theory of CE.

2.8 The Hydrothermal Systems of the Early Continental Crust

A third type of hydrothermal systems was proposed as cradle of life by U. Schreiber and coworkers (Schreiber et al. 2012, 2017, 2020; Mayer et al. 2015, 2017, 2018) over the past 15 years, namely hydrothermal systems located in flaws or cracks of the early continental crust. Those authors hypothesized that in geological fault zones, more or less vertical columns of water existed which might have reached a depth of several kilometers. Such hydrothermal systems include at least three gradients, first, increasing pressure. In a depth below 1 km, supercritical CO_2 might have been formed a separate nonpolar phase which might have enabled formation and storage of nonpolar organic compounds such as aliphatic alcohols or carboxylic acids. Provided that these monofunctional alkanes had a linear structure (*n*-alkyl chains), their formation has in turn, favored formation of lipid membranes. A second gradient concerns the temperature, which might have increased from 3 to 5 °C on the surface up to 50 or even 60 °C in a depth of 7 km and beyond. Ulrich et al.

assumed 2.76 that the higher radioactivity resulting from the decay of unstable isotopes caused a higher temperature in the lower levels of the continental crust than that known today. The third gradient or variable concerns the pore size along the interior rock walls of the fault zone.

Hydrothermal systems in fault zones of the continental crust might have worked as multifunctional reactors for the synthesis of organic compounds that finally enabled a molecular evolution is certainly an attractive concept which is clearly superior to that of the "black smokers" and "white smokers" because it avoids extreme temperatures and extreme pH values.

Mayer, Schreiber et al. (2020) also considered the problem of dilution of biomonomers in the primordial soup which hinders efficient formation of polypeptides/proteins and polynucleotides via polycondensation. Those authors developed a mathematical model of their hydrothermal systems which predicted spontaneous formation of vesicles at a depth around 1 km. These vesicles formed, in turn, a selective trap for amphiphilic molecules, especially for peptides composed of hydrophilic and hydrophobic prAAs in a suitable sequence. Further possible mechanisms in vesicle membranes were discussed. As evidence for a periodic vesicle formation in tectonic fault zones, the following experiments were reported. A commercial amphiphilic lipid (phosphatidylcholine) was subject to repeated pressure changes in mixtures of water, alcohol, and CO_2, so that states with subcritical and supercritical CO_2 followed each other in a timely sequence. Water droplets coated with lipid monolayers and vesicles based on lipid double layers were formed and transformed in each other following the rhythm of pressure changes. These results were considered to confirm the hypothesis of vesicle formation in hypothermal systems subject to pressure changes (tidal changes, volcanisms, etc.).

The crucial question is then, of course, what kind of organic compounds is formed in the hypothermal systems. Schreiber and Mayer gave two answers. They cited a publication of Fujioka et al. (2009) in which those authors report on prAA syntheses in a supercritical CO_2-water system mimicking a prebiotic soda fountain. Glyoxylic acid or pyruvic acid and hydroxylamine were used as starting materials at 31 °C or 60 °C. Gly and Ala were detected and small amounts of Asp and Arg. Furthermore, oligoAla's were found, but surprisingly no oligoGly's. The existence of significant amounts of hydroxylamine on the early Earth is highly questionable. Also questionable is the quality of that paper, because it contains strange comments: "Additionally, ½-cystine (1.2 mmol) and leucine (0.37 mmol) was detected at 60 °C and 31 °C, respectively, the sulfur atoms of leucine and ½-cystine may be derived from impurities of the product of hydroxylamine hydrochloride." Furthermore, the question, if non-protein AAs were formed, was not studied, and the questions, how hydroxylamine or the oximes of the keto acids was reduced under the mild conditions without catalyst remained open.

The second answer of Schreiber et al. is a paper describing the detection of organic compounds in fluid inclusions of Archean quartz, considered to be an analog of prebiotic chemistry on early Earth. This Archean quartz was found in the Jack Hills region of Western Australia and was believed to be older than three million years and, thus, acceptable as models of relicts of Archean hypothermal systems.

Table 2.2 Organic compounds detected in Archean quartz

Name	Formula	Name	Formula
Ethane	CH_3–CH_3	1,1,2,2-Tetrachloroethane	Cl_2CH–$CHCl_2$
Ethene	CH_2=CH_2	1,1,2,3-Tetrachloropropane	$Cl_2CHCHClCH_2Cl$
Ethyne (Acetylene)	HC=CH	2,2-dichloroethanol	Cl_2CH–CH_2OH
4-Methyl-2-pentanol	$(CH_3)_2CH$–$CH_2CH(OH)$-CH_3	3-Chloro-1-propanol	$ClCH_2CH_2CH_2OH$
3-Methyl-4-penten-1-ol	CH_2=$CHCH(CH_3)CH_2CH_2OH$	3,3-Dichloro-1-propanol	Cl_2CH–CH_2CH_2OH
3-Hexen-1-ol	CH_3CH_2CH=$CHCH_2CH_2OH$	1,3-Dichloro-2-propanone	$ClCH_2$–CO–CH_2Cl
Heptanal	C_6H_{13}–CHO	2,2,3-Trichloropropional	$ClCH_2C(Cl)_2CHO$

The detection of organic compounds in fluid inclusions of that quartz is certainly an important result, but does not necessarily indicate a cradle of chemical evolution. Numerous organic compounds were identified by chromatographic methods, a representative number of which is listed in Table 2.2. Four results are conspicuous. First, simple alkanoic acids, dicarboxylic acids, hydroxy acids, keto acids, and amino acids are absent. Second, nucleobases or nucleosides are absent. Third, alcohols having a long linear alkane chain (needed for formation of lipid layers) are absent. Fourth numerous chlorinated aliphatic compounds were detected. Three of the identified Cl-species (2,2-dichloroethanol, 1,3-dichloro-2-propanone, and 2,2,3-dichloropropional) are highly reactive alkylating reagents which will react with amino groups and SH groups and thus hinder formation of biologically active peptides and proteins (regardless of their synthetic pathway). In summary, these results indicate a chemistry, which is in contradiction to the requirements of a chemical evolution. Hence, any reliable evidence for the formation of "biomolecules" relevant for the success of chemical evolution in hydrothermal systems is lacking at the current state of affairs (2022).

2.9 The Ice–Water Interface as Cradle of Life

Over the past 20 years, H. Trinks and C. K. Biebricher have postulated that the ice/water interface might have been the birthplace of chemical evolution at temperatures below 0 °C. This scenario has important advantages over the aforementioned hypotheses of "black smokers" and "white-smokers" due to the low temperature and less extreme pH values. The rates of racemization are reduced by several orders of magnitude (s. Chap. 10), the risk of hydrolysis of RNA, the risk of hydrolysis of the amide groups in the side chains of Asn and Gln, and the risk of cleavage of the S–S bond of cysteine. However, these advantages are counterbalanced by a severe short-coming, namely by the fact that efficient syntheses of biomolecules, such as prAAs or nucleosides from primitive precursor molecules such as CO_2, CH_4, N_2, and NH_3 cannot have occurred under those conditions. There origin remains

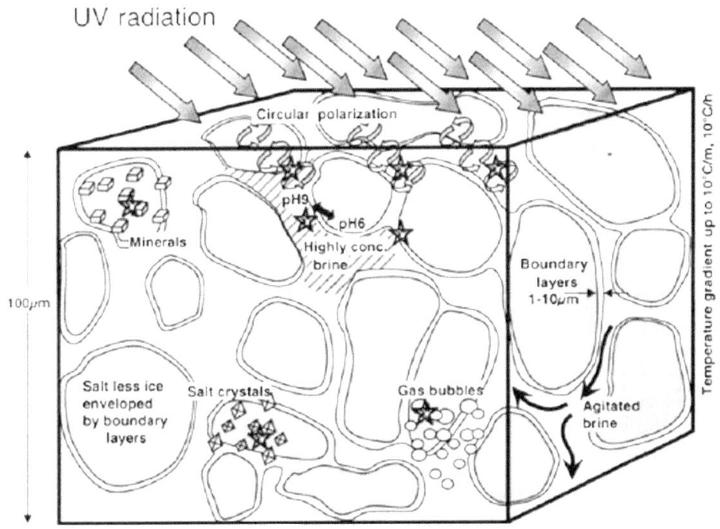

Fig. 2.3 Schematic structure of compartments at the boundary of sea-ice and sea water (Fig. 1 of Trinks et al. (2005) (with permission of Springer Berlin)

unexplained, and a continuous transport from other (warmer) places of the surface of Earth to regions having sea ice in a favorable temperature range is required throughout the year.

The freezing of sea water is a complex process, which can generate a variety of pores, compartments, and capillaries (Trinks 2001; Trinks et al. 2003, 2005). The complexity of this system may be enhanced by the presence of air bubbles and by the formation of precipitates that form salts and/or organic compounds (s. Fig. 2.3). Based on microscopic observations, Trinks et al. (2005) estimated that 1 m^3 of sea ice might contain 10^{13}–10^{14} compartments that together with the capillaries might possess a surface of 10^5–10^6 square meters. The existence of proton gradients in the direct neighborhood of the pores and capillaries, postulated by Lane (2015) as essential component for the birth of the first cell does also exist in the ice/water interface. Gradients in the range of pH 5–9 were reported by Bronshteyn and Chenov (1991). Furthermore, gradients and differences of the electric potentials up to 50 mV were detected (Steponkus et al. 1984). The reversible concentration and precipitation of salts upon freezing of water play an important role for both kinds of gradients. Furthermore, chromatographic effects were observed when mixtures of organic compounds were spread on a sea ice surface. The diffusion into pores and capillaries was accompanied by full or partial separation of the components. It is important for the proper understanding of all these effects that the seasonal up and down of the temperature causes a permanent rhythmic variation of all physical, chemical, and biological processes in the ice/water interface.

The hypothesis that the ice/water interface might have been the cradle of chemical evolution and life is supported by numerous observations. For instance, four research groups succeeded to synthesize RNA fragments at temperatures down to −18 °C. Stribling and Miller reported as early as in 1991 that they synthesized oligonucleotides up to DPs around 12–15 by polycondensation of nucleoside imidazolides (for details s. Sect. 7.3), when reaction times up to 8 weeks were used (Stribling and Miller 1991). Monnard et al. (2002, 2003) reported that copolycondensation of the four nucleoside imidazolides typical for RNA resulted in equimolar incorporation of pyrimidine and purine bases even at −18 °C. Vlassov et al. (2004) succeeded to connect RNA fragments by means of ribozymes in a partially frozen salt solution. Trinks et al. (2005) subjected a ATP solution to heating/cooling cycles between −7 and −24 °C in the presence of oligo(U) matrices. After 1 year, high molecular weight poly(A) was isolated. Of particular importance are the experiments of the Hollinger group (Attwater et al. 2010, 2013) who studied the self-replication of certain artificial ribozymes (catalytically active RNAs, s. Sect. 8.2) Those authors found that a temperature around 0 °C is more favorable for efficient self replication than temperatures >25 °C, because the base pairing between monomers and ribozyme templates is more stable and because the polycondensation involves less defects, when the temperature is lower. Finally, it is worth mentioning that numerous microorganism lives in the ice/water interface, the enzymes of which are still active at temperatures down to −15 °C. (Groudieva et al. 2004; Trinks et al. 2005).

When the "low temperature evolution" is compared with the "high temperature evolution" discussed for the "white smokers," two aspects are particular conspicuous:

1. Since the "white smokers" evolve CO_2 and H_2 in the presence of iron and nickel sulfides that are active redox catalysts, this scenario may develop its own production of basic chemicals, such as formaldehyde, formic acid, acetic acid, etc. Furthermore, the first cells and their precursors may gain chemical energy on the basis of redox processes. This potential is lacking in the ice - water system and all organic chemicals needed for the chemical evolution must be supplied from somewhere else.
2. The rates of racemization of prAAs and oligopeptides are extremely fast in the "white smoker" system with half-life times of few hours (s. Sect. 10.3). Since the temperature of the ice world is approximately 80 °C lower and the pH's close to 7–8, the rates of racemization will be lower by a factor of 10^8–10^{10}, a huge difference which is decisive for emerging and stabilization of a homochiral chemistry.

A review article published in 2012 Menor-Salvan et al. discussed various aspects of "Prebiotic chemistry in eutectic solutions at the water-ice matrix" (Menor-Salvan and Marin-Yaseli 2012).

Considering all the arguments presented in the previous sections, the author does not dare to draw a final decision, which one of the mentioned scenarios merits the title "birth place of chemical evolution." Regardless, if inorganic or organic compartments, low or high temperature environments merit this title, the combination of the RNA world with one of these scenarios is certainly the most promising concept for the elaboration of a convincing theory of chemical evolution.

References

Abdemale K, Engelhart A, Szostak J (2016) Nat Commun 7:11041
Abdemale K, Szostak J (2013a) Nat Chem 5(6):495
Abdemale K, Szostak J (2013b) Science 342:1098
Allen F (1899) Proc Birm Nat Hist Phil Soc 1:44–67
Altman S, Baer M, Guerrier-Taöada C, Vioque A (1986) TIBS 11:515–518
Apel C, Deamer D (2005) Orig Life Evol Biosph 15:323
Attwater J, Wochner A, Hollinger P (2013) Nat Chem 5:1011
Attwater J, Wochner A, Pinheiro V, Couison A, Hollinger P (2010) Nat Commun 1:76
Baly E, Heilbron J, Barker W (1921) J Chem Soc 119:1025–1035
Baly E, Heilbron J, Hudson P (1922) J Chem Soc 121:1075–1088
Bequerel P (1924) Bull Soc Astr Fr 98:393–417
Bernal J (1951) The physical basis of life. Routledge and Kegan Paul, London
Bernal J (1967) The origin of life. World Publishing, Cleveland
Black R, Bloser M, Stottrup B, Tavakley R, Deamer D, Keller S (2013) Proc Natl Acad Sci USA 110:13272
Blain J, Szostak J (2014) Annu Rev Biochem 83:615
Blocher M, Liu D, Luisi P (2000) Macromolecules 33:5787
Blocher M, Liu D, Walde P, Luisi P (1999) Macromolecules 32:7352
Böhler C, Hill A, Orgel L (1996) Orig Life Evol Biosph 26:1
Böhler C, Nielsen P, Orgel L (1995) Nature 376:578
Bronshteyn V, Chenov A (1991) J Cryst Growth 112:129
Budin I, Bruckner R, Szostak J (2009) J Am Chem Soc 131:9628
Budin I, Debnath A, Szostak J (2012) J Am Chem Soc 134:20812
Budin I, Prywes N, Zhang N, Szostak J (2014) Biophys J 107:1582
Budin I, Szostak J (2011) Proc Natl Acad Sci USA 101:1965
Bungenberg de Jong H (1932) Protoplasma 15:110–173
Cairns-Smith A (1982) Genetic take-over. Cambridge Univ. Press, Cambridge
Cedrangolo F (1959) The problem of the origin of the proteins in the origin of life on the Earth. Pergamon Press, London
Chen I, Roberts R, Szostak J (2009) Science 305:1474
Chen I, Saleh-Ashtani K, Szostak J (2005) J Am Chem Soc 127:13213
Chen I, Szostak J (2006) Chem Biol 13:139
Chen I, Szostak J (2009a) Proc Natl Acad Sci USA 102:7965
Chen I, Szostak J (2009b) Biophys J 87:988
Cody G et al (2000) Science 289:1337
Czech T (1987) Science 236:1532–1539
Czech T (1993) Gene 134:33–36
Deamer D (1985) Nature 312:792–794
Deamer D (2005) A giant step to artificial life? Trends Biotechnol 23:33
DeDuve C (1991) Blueprint of a cell: the nature and the origin of life. Carolina Biological Supply Company
DeDuve C (1998) Clues from present day biology: the thioester world. In: Brack A (ed) The molecular origin of life. Cambridge Univ. Press
DeDuve C, Miller S (1991) Proc Natl Acad Sci USA 88:10014
Diedrichsen U (1996) Angew Chem Int Ed 35:445
Diedrichsen U (1997) Angew Chem Int Ed 36:1886
Dyson F (1985) Origins of life. Cambridge Univ. Press, Cambridge

Egholm M, Burchhardt O, Chriszensen L, Behrens C, Frfeier S, Driver D, Berg R, Kim S, Norden B, Nelson P (1993) Nature 365:566
Ferris J (1993) Orig Life Evol Biosph 23:307
Fox S (1959) Naturwissenschaften 56:1–9
Fox S, Dose K (1977) Molecular evolution and the origin of life. Revised edn. Marcel Dekker, New York
Fox S, Harada K, Kendrick J (1959) Science 129:1221
Früh-Green G, Connolly J, Kelly D, Plas A, Grobery B (2004) AGU Geophys Monogr 144:119
Fujikawa S, Chen I, Szostack J (2005) Langmuir 21:12124
Fujioka K, Futamura Y, Shiohara T, Hoshimo A, Kanaya F, Manome Y, Yamamolto K (2009) Int J Mol Sci 10:2722–2732
Gandiner K, March T, Pace N (1985) J Biol Chem 260:5415–5419
Ganti T (1971) Azelet pricipuma (the principles of life). Gondolat, Budapest
Ganti T (1974) Biologia 22:17
Ganti T (1979) Biologia 27:161
Ganti T (2003) The principles of life. Oxford Univ. Press, Oxford
Gesteland R, Atkins J (1993) The RNA world. Cold Spring Harbor Laboratory Press, Cold Spring Harbor
Gilbert W (1986) Nature 319:618
Groen J, Deamer D, Kros A, Ehrenfreund P (2012) Orig Life Evol Biosph 42:295
Groudieva T, Kambourowa M, Yusef H, Royter M, Grote R, Trinks H, Antankian G (2004) Extremophiles 8(6):475–488
Guerrier-Takada C, Altman S (1984) Science 223:285–286
Guerrier-Takada C, Gardiner K, March T, Pace N, Altman S (1983) Cell 35:849–857
Hafenbrandt D et al (1995) Tetrahedron Lett 36:5179
Haldane J (1929) The origin of life. The rationalist annual 3 (Reprint in Bernal J (1967) Weidenfeld & Nicholson, London, pp 242–249)
Hanczyc M, Fujikawa S, Szostak J (2003) Science 302:618
Hanczyc M, Mansy S, Szostak J (2007) Orig Life Evol Biosph 37:67
Hanczyc M, Szostak J (2004) Curr Opin Struct Biol 8:660
Hargreaves W, Deamer D (2007) Biochemistry 17:3759
Hargreaves WR, Deamer DW, (1978) Biochemistry, 17, 3759–3762
Hargreaves W, Mulvihill S, Deamer J (1977) Nature 266:78
Hazen R, Deamer D (2007) Orig Life Evol Biosph 37:145
Hilz T, Blocher M, Walter P, Luisi P (2001) Macromolecules 34:2443
Himbert S, Chapman M, Deamer D, Rheinstätter M (2016) Sci Rep 6:31285. https://doi.org/10.1038/srep31285
Holm N, Dumont M, Ivanson M, Konn C (2006) Geochem Trans 7:7
Horowitz N (1986) Energy flaw in biology. Academic Press, New York
Huber C, Eisenreich W, Hecht S, Wächtershäuser G (2003) Science 301:938
Huber C, Wächtershäuser G (1997) Science 276:245
Huber C, Wächtershäuser G (1998) Science 281:670
Ikehara K (2002) J Biosci 27:165
Ikehara K (2005) Chem Rec 5:1078
Ikehara K (2009) Int J Mol Sci 10:1525
Ikehara K (2014) Orig Life Evol Biosph 44:299
Ishikawa K, Sato K, Shima Y, Urabe I, Yomo T (2004) FEBS Lett 576:387–390
Johnston W, Umrau P, Lawrence M, Glaser M, Bartel D (2001) Science 292:1319–1325
Joyce G (2002) Nature 418:214
Joyce G, Orgel L (1993) The RNA world (Chapter 1). Cold Spring Harbor Laboratory Press, Cold Spring Harbor, p 1
Joyce G, Orgel L (2006) Gesteland R, Czech T, Atkins J (eds) The RNA world (Chapter 2). Cold Spring Harbor Laboratory Press, p 23
Joyce G, Schwarz A, Miller S, Orgel L (1987) Proc Natl Acad Sci USA 84:4398–4402

Joyce G, Visser G, VanBockelt C, VanBoon L, Orgel L, VanVestrem J (1984) Nature 310:602
Joyce GF (1989) Gene 82:83
Kanat N, Tobe S, Hill I, Szostak J (2015) Angew Chem Int Ed 54:11735
Koppitz M, Nielsen P, Orgel L (1998) J Am Chem Soc 120:4563–4569
Kozlow I, Orgel L, Nelsen P (2000) Angew Chem Int Ed 39:4292
Kricheldorf H (2014) Polycondensation – history and new results. Springer, Berlin
Kruger K, Grabowski P, Zaug A, Sands J, Gottschling D, Czech T (1982) Cell 31:145–157
Kuneida N, Okamoto K, Fuei N, Kinoshita M (1981) Macromol Rapid Commun 2:711
Lane N (2015) The vital question: energy, evolution and the origin of complex life. W.W. Norton & Co, New York, London
Lincoln T, Joyce G (2009) Science 323:1229–1233
Lipman C (1924) Sci Mon 14:357–367
Luisi P (2006) The emergence of life. Cambridge Univ. Press, Cambridge
Mansy S, Schrum J, Krishnamurthy M, Tobe S, Tresco D, Szostak J (2008) Nature 454:172
Many S, Szostak J (2008) Proc Natl Acad Sci USA 105:13351
Martin W, Baross J, Kelly D, Russell M (2008) Nat Rev Microbiol 6:805
Martin W, Russel M (2007) Philos Trans Royal Soc 362:1887
Matreux T, Aikkila P, Scheu B, Braun D, Mast C (2024) Nature 628:110
Mayer C, Schreiber U, Davila M (2015) Orig Life Evol Biosph 45:139–148
Mayer C, Schreiber U, Davila M (2017) Life (MDPI) 7:3
Mayer C, Schreiber U, Davila M, Schmitz O, Bronja A, Meyer M, Klein J, Meckelmann S (2018) Life 8:16–31
Meierhenrich U, Munoz-Caro M, Bredehöft J, Jesseberger E, Thiemann W (2004) Proc Natl Acad Sci USA 101:9182–9186
Menor-Salvan C, Marin-Yaseli M (2012) Chem Soc Rev 41:5404
Mittapali G, Reddy K, Xiong H, Munoz O, Han B, DeRicardi F, Krishnamurthy F, Eschenmoser A (2007a) Angew Chem Int Ed 46:2470
Mittapali M, Ossornio Y, Guerrero A, Reddy K, Krishnamurthy R, Eschenmoser A (2007b) Angew Chem Int Ed 46:2478
Monnard P, Karaviuti C, Deamer D (2003) J Am Chem Soc 125:13734
Monnard P, Appel CL, KanavasiutiC, Deamer DW (2002) Astrobiology 2:139
Moore B (1913) The origin and nature of life. The Home Univ. Library of Knowledge, Henry Holt, London
Moore B, Webster T (1913) Proc Royal Sci B 87:163–176
Morgan T (1926) The theory of the gene. Yale Univ. Press, New Haven
Morowitz H (1992) Beginning of cellular life. Yale Univ. Press, New Haven
Mulkidjanian A (2009) Biol Direct 4:26
Muller J (1929) The gene as the basis of life. In: Dagger B (ed) Proc. Int. Congress of Planet Sci. 1926, Menasha Wisconsin, George Banta, pp 899–921
Müller F, Escobar L, Xu F, WEegrzyn E, Nainyta M, Amatov T, Chan CY, Pichler A, Carell T (2022) Nature 605:279–287
Namani T, Deamer D (2008) Orig Life Evol Biosph 38:329
Nelson K, Levy M, Miller S (2000) Proc Natl Acad Sci USA 97:3868–3871
Nielsen P (1993) Orig Life Evol Biosph 23:323–327
Nishikawa N, Arai M, Ono M, Iroh I (1993) Chem Lett 12:2017–2020
Oparin A (1924) Proiskhozhdenie Zhizny. Izd Moskovskii Rabochii, Moscow (Reprint in transl.: Bernal (1967), pp 199–214)
Oparin A (1938) Origin of life. McMillan, Ottawa
Oparin A (1961) Life: its nature, origin and development. Oliver and Boyd, Edinburgh
Oparin IA (1957) The origin of life on the Earth. Report on the International Symposium 1957 (Moscow, 1957). The Publishing House of the Academy of Sciences of the UDSSR
Orgel L (1968) J Mol Evol 38:381
Orgel L (1986) J Theor Biol 123:127
Orgel L (1998) Trends Biochem Sci 23:491

Orgel L (2000) Proc Natl Acad Sci USA 97:12503
Orgel L (2003) Orig Life Evol Biosph 33:211
Orgel L (2004) Crit Rev Biochem Mol Biol 39:99
Peters S, Semenov D, Hochleitner R, Trapp O (2023) Nat Sci Rep 13:6843
Rao M, Eichberg J, Oro J (1987) J Mol Evol 25:1
Robertson M, Joyce G (2012) Cold Spring Harbor Perspect Biol 4(5):a003608
Russell M, Arndt N (2005) Biogeoscience 2:97
Russell M, Daniel R, Hall A, Sherrington J (1994) J Mol Evol 39:231
Russell M, Hall A, Martin W (2010) Geobiology 8:355
Saladino R, Botta G, Costanzo G, DiMauto E (2012) Chem Soc Rev 41:5526
Schmidt J, Nielsen P, Orgel L (1997) Nucl Acid Rev 120:4563–4569
Schöning K, Scholz P, Guntha S, Krishnamurthy R, Eschenmoser A (2000) Science 290:1347
Schreiber U, Locker-Grütjen O, Mayer C (2012) Orig Life Evol Biosph 42(1):47–54
Schreiber U, Mayer C (2020) The first cell. Springer Nature, Switzerland
Schreiber U, Mayer C, Schmitz O, Rosendahl P, Bronja A, Greule M, Keppler F, Mulder I, Sattler T, Schöfer H (2017) PLoS One 12:e0177570
Schrum JP, Zhu TF, Szostak JW (2010) Cold Spring Harbor Persp. Biol 2:9
Schulte M, Blake D, Hehler T, McCollom T (2006) Astrobiology 6:364
Shapiro R (1987) Origin – a sceptics guide to the creation of life on Earth. Bantam Books/Summit Books, Simon Schuster, New York
Shibata SA, Yamashita S, Itoh Y, Yamashita T (1986) Biochim Biophys Acta 854:147
Sleep N, Bird D, Pope E (2011) Philos Trans Biol Sci 366:2857
Sousa F, Thiergart T, Landau G, Nelson-Sathi S, Pereira I, Allen J, Lane N, Maria W (2013) Philos Trans Biol Sci 368:1
Steponkus P, Stout D, Wolle J, Lovelace R (1984) Cryo-Letters 5:343
Stribling P, Miller S (1991) J Mol Evol 32:289
Szostak J (2012) J Syst Chem 3:2
Trinks H (2001) Auf den Spuren des Lebens. Shaker, Aachen
Trinks H, Schroeder W, Biebricher C (2003) Eis und die Entstehund ds Lebens. Saker, Aachen
Trinks H, Schroeder W, Biebricher C (2005) Orig Life Evol Biosph 35:429
Troland L (1914) Monist 24:92–133
Unrau P, Bartel D (1998) Nature 395:260–263
Ura Y, Beierle J, Lehmann L, Orgel L, Ghandiri M (2009) Science 325:73
Urey H (1952) The planets. Yale Univ. Press, New Haven, CT
Vlassov A, Johnston B, Landweber L, Kazakov S (2004) Nucl Acid Res 32:2966
Wächtershäuser G (1988) Microbiol Rev 52:452
Wächtershäuser G (1994) Proc Natl Acad Sci USA 91:428
Wächtershäuser G (2003) Science 289:1307
Weber A (1981) J Mol Evol 17:103
Weber A (1984) Orig Life Evol Biosph 15:17
Weber A (1998) Orig Life Evol Biosph 28:259
Westheimer F (1987) Science 235:1173–1178
Woese C (1979) J Mol Evol 13:95
Woese (1967) The genetic code: the molecular basis for genetic expression, Harper & Deus, N. Y.
Yarus M (2006) The RNA world (Chapter 9). Cold Spring Harbor Library Press, p 205
Young R (1965) Morphology and chemistry of microspheres from proteinoids. In: Fox S (ed) The origin of prebiological systems. Academic Press, New York
Zaug A, Czech T (1986) Science 231:470
Zaug A, Grabowski P, Czech T (1983) Nature 391:578–583
Zhang I, Peritz A, Merggers E (2005) J Am Chem Soc 127:4174
Zhu T, Ademale K, Zhong N, Szostak J (2012) Proc Natl Acad Sci USA 109:9828
Zhu T, Szostak J (2009a) J Am Chem Soc 131:5705
Zhu T, Szostak J (2009b) PLoS One 4(4):e5009

The Polycondensation Processes of the Chemical Evolution

3

A mistake is helpful, as soon as it has been understood and revised.

M. Rommel

3.1 Polycondensation

According to J. P. Flory, Nobel Prize laureate in 1974, all kinds of polymerization may be subdivided into two categories: chain growth polymerizations (CGPs) and step growth polymerizations (SGPs). This subdivision is based on kinetic aspects of the polymerization processes and not on thermodynamic aspects. The SGPs may again be subdivided into two groups, namely polycondensation and polyaddition, which are quite different from the thermodynamic point of view. Most polycondensations are characterized by a reaction entropy around zero, so that even SGPs with a small negative reaction enthalpy allow for syntheses of high molar mass polycondensates. This feature of SGPs arises from the fact that each condensation step which connects two molecules involves liberation of a new molecule, so that the balance of translational motion is zero.

This thermodynamic advantage also exists for all polymerization processes yielding biopolymers in living organisms, because all syntheses of biopolymers involve an elimination (condensation) reaction in every growing step. Nonetheless, enzymatic syntheses of biopolymers do not obey the definition of SGPs, because the kinetic course, which is decisive for the definition, exactly obeys the definition of chain growth polymerization. Hence, this unusual type of polymerization is called "condensative chain growth polymerization" (CCP, s. below). A more detailed comment on this kind of polymerization will be presented below in connection with the polymerization of α-amino acid *N*-carboxy anhydrides (NCAs).

However, before enzymatic syntheses of biopolymers were "invented" by the CE, syntheses of biopolymers must have occurred, and those syntheses followed the

H. R. Kricheldorf, *Life by Chemical Evolution?*
https://doi.org/10.1007/978-3-662-70752-4_3

pattern of normal SGPs. The systematic study of SGPs, above all of polycondensations processes, began in the year 1927 initiated by W. H. Carothers in the research laboratories of E. I. Dupont and had the invention and production of nylon fibers as consequence (Carothers 1931; Kricheldorf 2014; Chap. 2). A young coworker of Carothers, J. P. Flory, was concerned with experimental and theoretical analyses of the polycondensations performed in the labs of E. I. Dupont. After the early death of Carothers in 1937, Flory started his academic career and elaborated the first theory of SPGs (Flory 1946, 1953), which still today, 70 years later, forms the basis of any text-book chapter on polycondensation. This classical theory of SGPs is characterized by the following three points:

1. Monomers, oligomers, and polymers can react with each other at any time and at any concentration.
2. Polycondensations differ from polyadditions by elimination of a by-product in every growing step, so that the level of translational mobility remains constant, and thus, the reaction entropy is close to zero (changes of rotational motions modify this simplified picture).
3. Cyclization reactions don't play any role.

Over the past 50 years, several research groups elaborated two more criteria for the definition of polycondensation.

1. When polycondensations are conducted in the melt or in solution, so that all chains maintain full mobility, cyclization reactions compete with chain growth at any time and at any concentration. In other words, point 3 of Flory's theory was revised (Kricheldorf and Schwarz 2003; Kricheldorf 2014; Chap. 6).
2. The molar concentration of all active species [La], monomers, oligomers, and polymers, decreases with increasing conversion according to the law of self-dilution (Eq. 3.1; Kricheldorf 2009). This self-dilution favors, in turn, cyclization at the expense of chain growth.

$$[\mathrm{La}]_t = [\mathrm{La}]_0 (1 - p) \tag{3.1}$$

where $[\mathrm{La}]_0$, $[\mathrm{La}]_t$ are molar concentration at the beginning or at a later time; p = conversion.

Flory formulated four simple mathematical equations as fundamental of his classical theory. Equation 3.2 correlates the average degree of polymerization (DP) with the progress of conversion.

$$\mathrm{DP} = 1/(1 - p) \tag{3.2}$$

The molar fraction of one sort of chains, n_x (defined by its degree of polymerization x) varies with the conversion according to Eq. 3.3:

$$n_x = (1 - p)p^{x-1} \tag{3.3}$$

The weight fraction, w_x, of chains having an identical degree of polymerization, x, can be calculated from Eq. 3.4

$$w_x = (1-p)^2 p^{x-1} \tag{3.4}$$

The dispersity of a virgin polycondensate, D, follows Eq. 3.5

$$D = \mathrm{Mw} / \mathrm{Mn} = 1 + p \tag{3.5}$$

where Mn and Mw are number and weight average molecular weight of the polycondensate.

Particularly important for the discussion of experimental results in this book are Flory's equation which correlate the DP with stoichiometric imbalance of two comonomers (e.g., nucleoside and phosphoric acid), case I. Furthermore, he calculated the influence of a monofunctional chain terminator on the DP (case II). For case I, the following definition was considered:

$$\mathrm{DP} = (1+r) / 2r(1-p) + 1 - r \tag{3.6}$$

In this equation, r is defined as N_{a0}/N_{b0}, the initial concentrations of functional groups of the monomers "a_2" and "b_2."

With 100% conversion, Eq. 3.6 degenerates to Eq. 3.7

$$\mathrm{DP} = (1+r) / (1-r) \tag{3.7}$$

This equation may also be applied to case II, i.e., for the presence of monofunctional chain terminators (ct), when r is defined according to Eq. 3.8:

$$r = N_{a0} / (N_{b0} + \mathrm{ct})$$

At this point, it should be emphasized that the DPs calculated according to Eqs. 3.6 and 3.7 represent the sum of a_2 and b_2 units, and thus, are twice as high as the number of repeat units (the original meaning of DP). Hence, it is necessary to half the calculated DPs, when polycondensations of "a-b" monomers, such as amino acids, are considered. A concrete example and for the following chapter relevant example should be given. When the number of monofunctional chain terminators (e.g., acetic acid or ammonia), N_{ct}, amounts to 2/3 relative to N_{a0} or N_{b0}, a DP of 4 will result. In the case of "a-b" monomers, N_{ct} = 2/3 of N_{ab} yields a DP of 2, meaning dimers. A DP of 2 will also result, when in the absence of chain terminators the conversion stops at 50%. The probabilities of the formation of higher oligomers are listed in Table 3.1. When equimolar amounts of carboxylic acids or amines are present in a pool of prAAs, the probability of the formation of polypeptides or proteins is almost zero regardless of the synthetic method. As will be shown in Chaps. 4 and 9, this is the most likely scenario found for model experiments and also found for the chemistry in space. The same conclusion can be drawn for any synthesis of polynucleotides from nucleosides when monofunctional alcohols or alkyl phosphates are present in the reaction mixture.

In the year 1972, two British research groups headed by the physicochemists R. F. T. Stepto (1972) and M. Gordon and Temple (1972a, b) published a mathematical modeling of the kinetic course of irreversible polycondensations. Four different mathematical approaches were compared, and nearly identical results

Table 3.1 Molar ratios (probabilities) of oligomers for various degrees of polymerizations (x) at a conversion of 50% (average DP(x) = 2)

x	N_x/N	x	N_x/N
1	1/2	8	(1/2$^{)8}$
2	$(1/2)^2$	9	$(1/2)^9$
3	$(1/2)_3$	10	$(1/2)^{10}$
4	$(1/2)^4$	11	$(1/2)^{11}$
5	$(1/2)^5$	12	$(1/2)^{12}$
6	$(1/2)^6$	20	$(1/3)^{20}$
7	$(1/2)^7$		

N = sum of all oligomers

Scheme 3.1 Competition between chain growth and cyclization in polycondensations in solution and in the melt

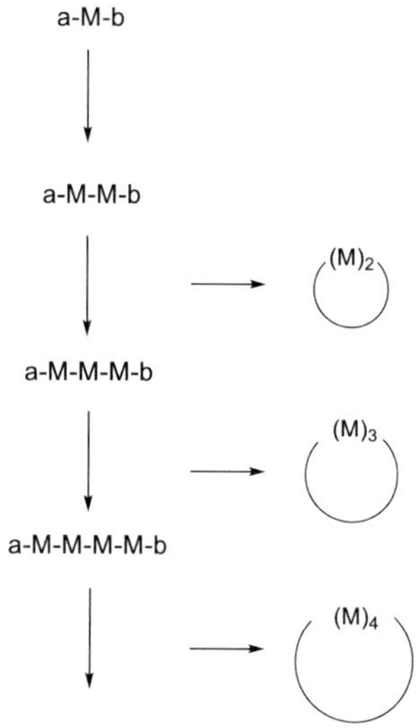

were obtained. They revised the classical theory of Flory with regard to the influence of cyclization reactions as follows:

1. Cyclization reactions compete with chain growth at any concentration and at any stage (conversion) of the entire process (s. Scheme 3.1).
2. The fraction of cycles increases with higher conversions and at 100% conversion all reaction products will be cycles.
3. Dilution favors cyclization at the expense of chain growth according to the Ruggli-Ziegler dilution principle.

Numerous experimental facts confirming that these calculations and conclusions were contributed by Kricheldorf and coworkers after 1999 by means of MALDI-. TOF mass spectrometry (Kricheldorf and Schwarz 2003; Kricheldorf 2014). The Ruggli-Ziegler (1934) dilution principle says that a dilution of the reactants by a factor of 2 enhances the rate of cyclization over the rate of chain growth (k_c/k_p) increases by factor of 2. As a consequence of this principle, the s. c. Carothers equation (Eq. 3.8) needs to be replaced by Eq. 3.9.

$$\mathrm{DP} = 1/(1-p) \tag{3.8}$$

$$\mathrm{DP} = 1/\left[1-p\left(1-X^{\alpha}\right)\right] \tag{3.9}$$

with $\alpha = k_c/k_p$.

What are the consequences of the polycondensation theories presented above for the hypothesis of chemical evolution? Two aspects are of paramount importance:

1. High yields of high molecular weight biopolymers can only be achieved in highly concentrated solution. The "hot dilute soup" of J. Haldane is exactly the contrary to what is needed for a successful polycondensation. Polycondensation in bulk, for example, by evaporation of an aqueous solution of biomonomers on the hot slopes of a volcano (s. Sect. 6.1) results in decomposition of saccharides, in particular, of ribose, and causes rapid racemization of prAAs (s. Sect. 10.3). Adsorption of monomers, oligomers, and polymers on the surfaces of minerals and polycondensations in canals or pores of minerals or ice are, in principle, more favorable. A strong fixation of oligomers on surfaces of minerals has the advantage that it may hinder cyclization, but the reduced mobility also hinders chain growth. However, a dynamic dissociation equilibrium with the solution enables cyclization again.
2. The presence of large amounts of monofunctional compounds such as carboxylic acids, aldehydes, alcohols, or amines (i.e., high N_{ct} values in Eq. 3.8) prevents any chain growth by blocking of reactive end groups. To best knowledge of the author, R. Shapiro (2006) is the only author, who has, at least shortly, mentioned this line of argumentation.

The above discussion necessarily implies that it does not suffice to demonstrate the possibility of a CE, when the formation of several prAAs or nucleosides is evidenced in model reactions. If formation of biomonomers is accompanied with formation of large amounts of by-products that can prevent any polycondensation of the biomonomers, the reaction conditions of the model experiment may per se be interesting, but they are useless as evidence for the origin of life on the basis of a CE. For the vast majority of model experiments, structure and quantity of functional by-products were not studied. To find realistic reaction conditions for the formation of biopolymers on the early Earth, identification and quantification of functional by-products are as important as the synthesis of the desired biomonomers itself. With exception of S. L. Miller, all authors of model syntheses have ignored this important aspect. The exploration of organic compounds delivered by carbonaceous meteorites to the Earth was far more accurate, comprehensive, and satisfactory in this regard (s. Chap. 9).

Finally a new mathematical simulation of polycondensations in small droplets should be mentioned. Szymanski and Sosnowski (2025) demonstrate that all step growth polymerization performed in small volumes deviate from the course of SGPs in larger volumes, so that formation of longer chains with lower dispersities is favored. This new insight means that polycondensation in small compartments was favorable for the prebiotic formation of biopolymers with narrow molecular weight distribution, regardless, if polypeptides or polynucleotides are concerned. For the sake of completion, it should be mentioned that experimental and theoretical studies of chemical reactions in mini-reactors are not absolutely new, and the reader interested in that field may consider the literature cited in the Szymanski-Sosnowskipaper.

3.2 Condensation Chain Polymerization (CCP)

As already mentioned above, CCP is a special version of chain polymerizations which is typical for the enzymatic syntheses of biopolymers and which is characterized by the following properties:

A. Each growing step involves an elimination reaction in contrast to all other chain polymerizations, and this elimination (condensation) step is the only analogy to polycondensation processes.
B. There is only one kind of growing steps, namely the reaction of the monomer with the active chain end.
C. A CCP may be initiated by an initiator forming a stable end group, and the chain growth proceeds solely on one chain end.
D. The degree of polymerization can be controlled via the monomer initiator ratio, provided that the initiation reaction is as fast or (better) faster than the propagation.
E. Frequently narrow molecular weight distribution with dispersities as low as 1.1 can be achieved.
F. There are no cyclization reactions that compete with the chain growth.
G. In the absence of termination steps, one chain end remains active even after 100% conversion of monomer A, and the polymerization may be continued by addition of monomer B. In this way, deblock copolymers can be synthesized. Repetition of this procedure with monomer A or monomer C allows for preparation of multiblock copolymers.

This list of characteristic properties correctly interpreted makes clear, why the biosynthesis of biopolymers is based on CCPs and not on polycondensations. Only a CCP allows for syntheses of copolymers with proper control of the chain length and with specific sequences of different monomeric units. A full control of all these structural parameters is, of course, an unavoidable prerequisite for a reproducible synthesis of proteins with well-defined biological functions. A more detailed comment on the aforementioned points A)-G) should follow here.

Ad A): The elimination steps entail the advantageous thermodynamic property that the propagation CCPs, in contrast to all other chain growth polymerizations, does not involve a loss of entropy ($\Delta S = 0$). Therefore, even CCPs involving a low reaction enthalpy only allow for the formation of high molar mass (bio)polymers. Even a reaction enthalpy around 0 does not hinder a successful chain growth, if the eliminated by-product is removed from the reaction equilibrium. This removal may happen in three ways:

1. By evaporation as it is typical for the elimination of CO_2.
2. Rapid crystallization from the reaction mixture as it is typical for the elimination of NaCl.
3. Chemical modification of the by-product. Neutralization of eliminated HCl or hydrolysis of eliminated pyrophosphate groups are typical examples.

This scenario has two important advantages for the course of the CE and for living cells. First, it was not necessary to synthesize highly energy rich monomers to perform successful syntheses of biopolymers. Second, the CE and living cells don't need to fight a significant loss of entropy, an advantage which is particularly important, because the synthesis of an ordered sequence of various monomers (prAAs or nucleotides) involves a slight loss of entropy anyway.

A, B, C, D, and G: These properties of a CCP are the prerequisites for the synthesis of a biopolymer starting out from an initiator (primer) with full control of chain length and sequence of monomers. A quite simple model of a biopolymer synthesis in a test tube is the polymerization of glucose-1-phosphate by means of the enzyme "potato-phosphorylase." The start of this polymerization results from addition of an oligo-amylose which acts as primer.

Ad E: Here it should be mentioned that polymer chemist was not capable so far to realize a polymerization process which yields chains of identical length. The many times repeated biosynthesis of identical enzyme molecules is an extraordinary performance of living cells, which has never been realized by polymer syntheses in laboratories or in technical production processes.

Ad F: The absence of competing cyclization reactions (the presence of which is characteristic for polycondensations) allows for the synthesis of a homogeneous, uniform topology of the resulting (bio)polymers. Living cells also produce cyclic biopolymers if needed, but their synthesis is based on a separate cyclization step after the complete synthesis of the linear precursor.

The characteristic differences between CCP and polycondensation are of outmost importance for the success of the CE and its elucidation and explanation. Even, when it becomes possible in the frame of an undirected evolution to demonstrate syntheses of biopolymers via polycondensation in model experiments, these polycondensations would not explain, how the first cell has learned to synthesize biopolymers via CCPs. With the formation of a RNA-based protein synthesis controlled by genetic code, this problem is solved, but the syntheses of the components of such a machinery also required the existence of CCPs. The auto-catalyzed self-reproduction of ribozymes, which was elaborated in the frame of the research on "evolution in a test tube," provides at least a partial response to this question (s. Sect. 8.2).

Interestingly, a model synthesis of (co)polypeptides having all the characteristics of a CCP has been found, namely certain polymerization procedures of α-amino acid *N*-carboxy-anhydrides (NCAs). In contrast to almost all books or review articles dealing with CE, P. Luisi published a short comment on polymerizations of

NCAs in his book *The emergence of life* (Luisi 2006). Unfortunately, he called the polymerizations of NCAs "polycondensations," and thus, demonstrated that he had not understood the difference between polycondensation and CCP. However, polymerizations of NCAs are not only of interest because as models of CCPs in the field of biopolymers and their analogs, several authors have also hypothesized that polymerizations of NCAs might have played an important role for the formation of protein precursors in an early stage of the CE. Since not only hypotheses but also experimental studies in this direction were published, the following section is dedicated to a more detailed discussion of the NCA chemistry.

3.3 Polymerization of α-Amino Acid *N*-Carboxy-anhydrides (NCAs)

NCAs are highly reactive monomers, which are sensitive to hydrolysis and sensitive to heat, and therefore, they are commercially not available. Due to their high reactivity, numerous nucleophilic and basic compounds can be used as initiators or catalysts of their polymerization. Under the conditions of the CE, only protic nucleophiles such as water, primary alcohols, and primary amines might have played the role of initiators. Initiation with primary aliphatic amines is somewhat faster than the chain growth and when the NCA/amine ratio is low, only oligomers can be formed. Initiation with primary aliphatic amines is several orders of magnitude faster than initiation with alcohols, so that in the presence of primary amines, other initiators have not played any role. Ammonia and the more reactive methylamine are the most likely candidates as initiators under prebiotic conditions. He somewhat simplified mechanism of an amine-initiated polymerization of NCAs illustrated in Scheme

Scheme 3.2 Mechanism of primary amine initiated polymerizations of α-amino acid NCAs

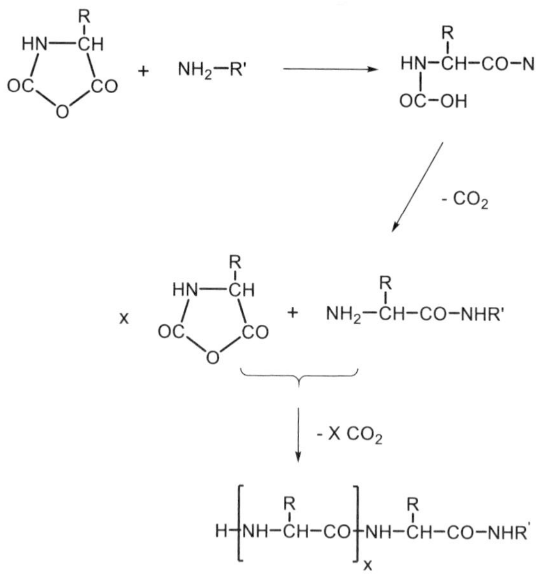

3.2. The carbamic acid end group which is formed as reaction intermediate is unstable under neutral or slightly acidic conditions. It is temporarily stabilized as carbamate anion under slightly alkaline conditions, but it can continue the propagation. A detailed discussion of polymerization mechanism of NCAs has been published by Kricheldorf (1987 and 2006).

The following properties of this polymerization process are characteristic for the course of a CCP:

(a) One dead end group is formed by incorporation of the initiator, and one end group remains active (the NH_2 group).
(b) The chain growth proceeds exclusively via the reaction of NCAs with the active (amino) end group
(c) Provided that the resulting polypeptides stays in solution, a narrow molecular weight distribution will be achieved.
(d) After complete consumption of the first NCA, the polymerization may be continued by addition of a second NCA, so that a two-block copolymer will be formed.
(e) No cyclization reactions occur.

At this point, the question arises, if the amine (or water)-initiated polymerization is of interest only as model system of a polypeptide synthesis via CCP, or if it may have played a realistic role in the CE. Any plausible answer on this question requires at first an answer on the question, if and how the NCAs of prAAs may have been formed under prebiotic conditions. Those methods used for syntheses of NCAs in laboratories are not acceptable as models of NCA syntheses under prebiotic conditions. However, A, Comeyras and coworkers found a synthetic approach, the reaction conditions of which come close to the requirements of a prebiotic NCA synthesis (Taillades et al. 1998, 1999; Comeyras et al. 2002). As outlined in Scheme 3.3, the first step of the Comeyras approach is formation of urea derivatives of prAAs (s. c. hydantoic acids). (Iso)cyanate ion was certainly present in the primordial soup. The urea groups react with gas mixtures containing NO and O_2 and yield the desired NCAs via instable hydroxy diazonium groups. This reaction pathway is

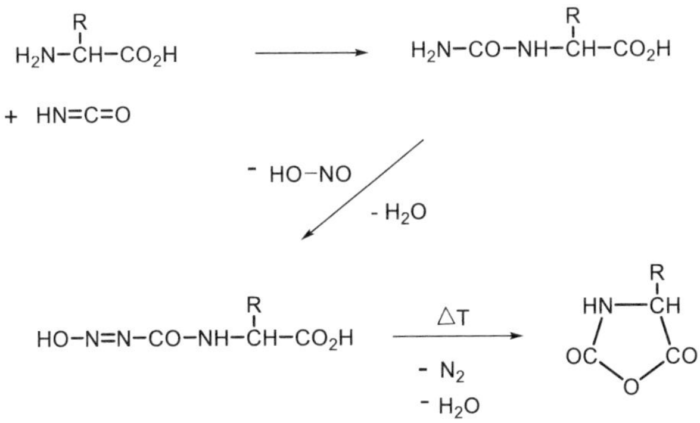

Scheme 3.3 NCA synthesis by A. Comeyras and coworkers

successful even in the presence of cold water, although NCs are quite sensitive to hydrolysis and Comeyras and coworkers were able to isolate several NCAs in this way by rapid separation from the aqueous reaction mixture. The nature of the final stable reaction products depends mainly on the pH. In acidic water, protonate prAAs are the final products, whereas under neutral or slightly basic conditions, oligo- and polypeptides are the stable end products.

Furthermore, there are several publications reporting on syntheses of oligopeptides via an intermediate formation of NCAs that were not isolated. This scenario is characteristic for (formal) polycondensations of prAA derivatives having an activated CO group, when these prAA derivatives are exposed to a carbonate buffer (Scheme 3.4, upper reaction sequence). The first experiments suggesting an intermediate formation of NCAs were published by Korshak et al. (1954), who introduced CO_2 into aqueous solutions of Ala or Gly ethyl esters at 50 °C. Without introduction of CO_2, no oligopeptides were formed and cyclic dipeptides (2,5-Dioxopiperazines) were the only reaction products resulting from the cyclization of the intermediately formed dipeptide esters (Kozarenkto et al. 1957, Scheme 3.4). Analogous results were reported by A. Brack (1982, 1987), with prAA 44nitrophenyl esters or S-ethyl esters. The polycondensations of prAA adenylates described by M. Paecht-Horowitz will discussed in Chap. 6 in more detail. This is also true for the polycondensations of prAAs bay means of carbonyl diimidazole, which were extensively studied by L. Orgel and coworkers. Both research groups considered the intermediate formation of NCAs as most likely reaction pathway. Finally, the work of Leman et al. (2004) is worth mentioning. Those authors studied the polycondensation of prAAs activated by COS and also proposed an intermediate formation of NCAs. A review article summarizing hypotheses on the role of NCAs and other peptide-forming methods in the prebiotic evolution was published by the Commeyras group in 2005 (Pascal et al. 2005).

Scheme 3.4 Different reactions of α-amino acid esters and dipeptide esters

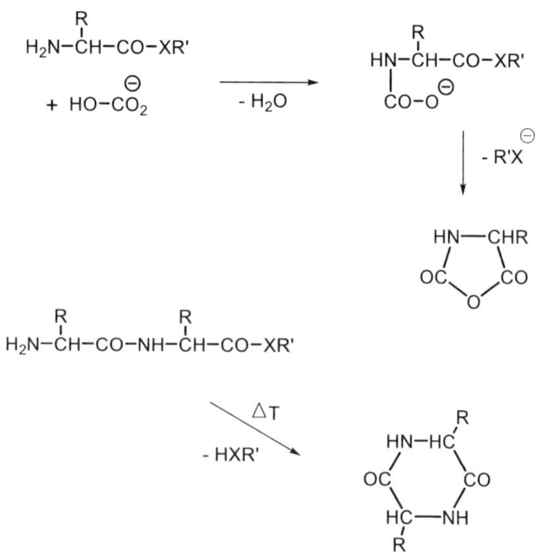

To what extent are all these results useful as arguments in favor of a prebiotic synthesis of proteins? At first, one has to consider that the "Comeyras approach" to the synthesis of NCAs requires a high concentration of NO and O_2 in the atmosphere, because the freshly prepared NCAs rapidly react with water and/or amines. As discussed in Chap. 2, such a strongly oxidizing composition of the early atmosphere cannot have existed. The composition of the atmosphere to day containing 20% of oxygen is the result of the photolysis reaction occurring in green plants, but an efficient, oxygen producing process exclusively based on inorganic chemical did certainly not exist on the early Earth. Furthermore, even in the today atmosphere almost completely consisting of nitrogen and oxygen only traces of NO exist, and these traces mainly result from diesel engines. Moreover, all NCAs decompose above their melting point, typically above 60 °C, they are highly sensitive to hydrolysis and they are extremely sensitive to racemization. Hence, NCAs cannot have played any role in scenarios requiring high temperatures and slightly alkaline conditions. They might have been formed and survived for hours at temperatures below 5 °C (e.g., in the ice/water world of H. Trinks, s. Chap. 2), but there is another argument against an important role of NCAs in the prebiotic formation of proteins.

This counterargument consist of the fact that the NCAs of most trifunctional prAAs such as Ser, Thr, Cys, His, Lys, and Arg) are instable even around 20 °C and undergo a rapid rearrangement as outlined for selected examples in Scheme 3.5 (Saito 1964; Hirschmann et al. 1971). In the case of Glu-NCA, the rearrangement yielding pyroglutamic acid requires temperatures above 20 °C. As reported by K. W. Ehlers et al. (1977), the instability of their NCAs prevented the (co)

Scheme 3.5 Rearrangement reactions of NCAs of trifunctional α-amino acids

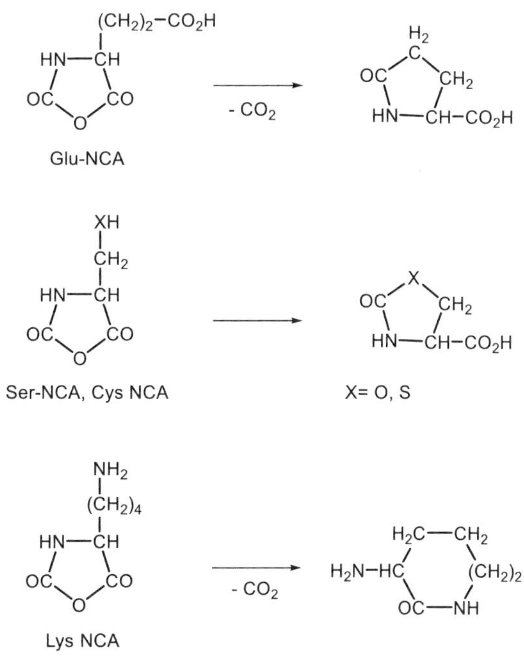

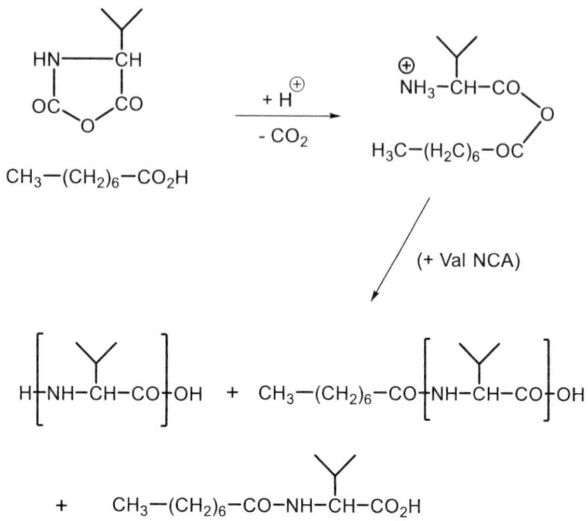

Scheme 3.6 Reaction products formed from val-NCA and octanoic acid when heated

polymerization of Ser, Thr, and His by means of carbonyl diimidazole. Yet without incorporation of those prAAs, a CE of hydrophilic proteins cannot have taken place. Yet, peptides exclusively composed of hydrophobic prAAs turn insoluble in water above a DP around 5.

Furthermore, significant amounts of NH_3 and methylamine must have been absent, because their rapid reaction with NCAs (within minutes or hours depending on the temperature) would have prevented the formation of longer peptide chains. However, at least the intermediate formation of large amounts of ammonia is required to explain the formation of prAAs and other amino and imino acids under prebiotic conditions, as demonstrated in numerous model experiments (s. Chap. 4) in agreement with the extracts of meteorites. It is, of course pretty naïve to believe that NH_3 and methylamine were present in an early stage of the CE, but disappeared complete after sufficient formation of various amino acids to enable syntheses of polypeptides and finally of proteins. Moreover, one has to consider that an excess of simple carboxylic acids, as it was found in the experiments of S. L. Miller (Chap. 4) and in the extracts of meteorites (Chap. 9), may also have hindered a successful polymerization of polypeptides. For example, J. Szostak and coworkers (Izgu et al. 2016) have demonstrated that heating of Val-NCA in the presence of oleic acid or octanoic acid yields *N*-acyl Val and *N*-acyl oligopeptides. A mixed anhydride was identified as reaction intermediate (Scheme 3.6). Again one has to believe in a miraculous disappearance of all the initially formed carboxylic acids prior to formation and polymerization of NCAs, or these polymerizations have never taken place.

In summary, all the experimental facts presented above allow for the conclusion that the NCAs cannot have played any significant role for the formation of polypeptides and proteins under prebiotic conditions.

References

Brack A (1982) Biospheres 15:201
Brack A (1987) Orig Life 17:367
Carothers W (1931) Chem Rev 8:353
Comeyras A, Collet H, Moiteau L, Taillades J, Vandanabeele-Trambonze O, Biron J, Plasson R, Miron L, Lagrille O, Martin H, Seilsis F, Dobrijevic M (2002) Polym Int 51:661
Ehlers K, Girard E, Orgel L (1977) Biochem Biophys Acta 491:253
Flory J (1946) Chem Rev 39:137
Flory J (1953) Principles of polymer chemistry, Chapters III and VIII. Cornell Univ. Press, Ithaca, NY
Gordon M, Temple W (1972a) Makromol Chem 152:277
Gordon M, Temple W (1972b) Makromol Chem 160:263
Hirschmann R, Schwam H, Strachan R, Schoenewald E, Barkemeyer A, Miller S, Con J, Garsby V, Veber D, Denkewalter R (1971) J Am Chem Soc 93:2746
Izgu E, Börkholm A, Kamat N, Lebfeld V, Zhng W, Jia T, Szostak J (2016) J Am Chem Soc 138:16669
Korshak V, Poroshin K, Kozarenko T (1954) Izv. Akad. Nauk USSR Ordel Khim Nauk, p 663
Kozarenkto T, Poroshin K, Kuruyin K (1957) Izv. Akad. Nauk, p 640
Kricheldorf H (2006) Angew Chem Int Ed 45:5752
Kricheldorf H (2009) Macromol Rapid Commun 29:1695
Kricheldorf H (2014) Polycondensation – history and new results. Springer, Berlin
Kricheldorf H (1987) α-Amino Acid N-Carboxyanhydrides and Related Heterocycles, Springer, Berlin-Heidelberg
Kricheldorf H, Schwarz G (2003) Macromol Rapid Commun 24:359
Leman I, Orgel L, Ghadiri M (2004) Science 306:283
Luisi P (2006) The emergence of life. Cambridge Univ. Press, Cambridge
Pascal R, Boiteau L, Comeyras A (2005) Top Curr Chem 259:69
Saito T (1964) Bull Chem Soc Jpn 37:624
Shapiro R (2006) Quart Rev Biol 51:105
Stepto R (1972) Makromol Chem 152:263
Szymanski R, Sosnowski S (2025) Chem Eng Sci 301:120655
Taillades J, Beuzelin I, Garrel L, Tobavcizky V, Bried C, Comeyras A (1998) Orig Life Evol Biosph 28:61
Taillades J, Collet H, Garrel L, Beuzelin I, Boiteau L, Choukrun H, Comeyras A (1999) J Mol Evol 48:638
Ziegler K (1934) Ber Dtsch Chem Ges 67B:139

Model Syntheses of Amino Acids

4

Not all molecules that can be synthesized are useful and not all molecules that may be useful can be synthesized.

H. R. Kricheldorf

4.1 The Amino Acid Syntheses of Stanley L. Miller and Coworkers

Peptides and proteins are the most diversified and abundant class of "biochemicals" which exist in animals and in the human body. Furthermore, all biocatalysts that are active in plants contain (poly)peptide chains. Hence, the origin of peptides and proteins is of paramount importance for any theory about the emergence of life. The existence of protein α-amino acids (prAAs) on the early Earth may be explained by two quite different hypotheses. Hypothesis I assumes that all prAAs originate from space delivered to the Earth by comets, meteorites and, above all, by large amounts of micrometeorites (s. Chap. 9). The alternative hypothesis II says that the prAAs were formed on Earth by efficient reaction pathways starting out from simple chemicals, such as ammonia, aldehydes, cyanide ions, and water. These different origins are not necessarily alternatives, but might have coexisted and even had a synergistic interaction, when optically active AAs imported from space (s. Chap. 9) have favored a catalytic formation of prAAs having an excess of one enantiomer (s. Chap. 10).

The history of model syntheses of prAAs did not begin with the famous experiments of S. L. Miller (as usually stated in the literature), but with the work of Löb in 1913 (Löb 1913). Furthermore, Dar and Mukerjee conducted similar experiments as did Miller at the same time (1954). Löb exposed gas mixtures consisting of CO_2, H_2O, and NH_3 to electric discharges and observed formation of glycine. Maybe, other amino acids, such as Ala and ß-alanine, were also present in his reaction mixtures, but the lack of sensitive analytical methods prevented their detection. Dar and

H. R. Kricheldorf, *Life by Chemical Evolution?*
https://doi.org/10.1007/978-3-662-70752-4_4

Mukerjee irradiated mixtures of aliphatic diols or saccharides and nitrates of sodium, potassium, and ammonium with sunlight and found a rapid formation of amino acids, but they were not able to identify individual prAAs. The existence of large amounts of nitrates on the early Earth is unlikely, but this is also true for the reducing atmosphere used by Miller.

After a short communication in Science 1953, Miller published a detailed description of his experiments in 1955 (Miller 1953, 1955) (Table 4.1). Following advices of his mentor Urey, Miller used a reducing atmosphere consisting of CH_4, NH_3, H_2, and H_2O. The man reaction product of all experiments was an insoluble polymeric mass which was not characterized in detail. Miller needed several variations of the reaction conditions until he succeeded with the identification of a few prAAs. For the first fractionation of the crude water-soluble reaction products, he used column chromatography equipped with ion-exchange resins, and in this way, he was able to isolate Gly ad Ala. Using paper chromatography, let him conclude that Asp, Glu, and Val were also present in minor amounts. Particular important but ignored by most reviewers citing his work was the finding that α-Abu, ß-Ala and Sar were also formed in sizable amounts. Miller described three somewhat different procedures (runs I-III) and mentioned that the data of run III were the average of four similar experiments. This run is of extraordinary interest because Sar proved to be the most abundant amino (imino) acid and not a prAA. This result demonstrates that arbitrarily selected prebiotic conditions do not necessarily yield prAAs as main reaction products, a result which has never been mentioned in the numerous reviews and books on chemical evolution presenting hymns about the work of Miller.

Furthermore, Miller isolated three aliphatic carboxylic acids and two hydroxy acids, the molar amount of which exceeded that of the prAAs by a factor of 3. Weakly basic products and peaks of numerous unidentified compounds were also

Table 4.1 Syntheses of prAAs by means of electric discharges in a reducing atmosphere

Gas mixtures	prAAs	Non-prAAs	By-products	References
CH_4, NH_3, H_2O, H_2	Alaa, Gly, (Asp, Glu, Val)	α-Abu, ß-Ala, Sar	Formic a., acetic a., propionic a., glycolic a., lactic a., amines	Miller (1955)
CH_4, NH_3, H_2O, H_2	Ala, Gly, Asp, Glu	α-Abu, ß-Ala, Sar. *N*-Methyl-Ala, iminodiacetic a., iminoacetic-propionic a.	Formic a., acetic a., propionic a., glycolic a., lactic a., α-hydroxybutyric a., urea, *N*-methyl urea	Miller (1957)
CH_4, NH_3, H_2O, H_2 or CO/CO_2, N_2, H_2O, H_2	Ala, Gly, Asp, Glu, Ser, Val	No information	No information	Schlesinger and Miller (1983)
CH_4, NH_3, H_2S, CO_2	Ala, Gly, Asp, Glu, Ile, Leu, Ser, Thr, Val	α-Abu, γ-Abu, ß-Ala, isoSer, isoVal, S-methyl-Cys and its sulfoxide and sulfone, homocysteine, ethionine	Methylamine, ethylamine, ethanol, cysteamine, no information about carboxylic acids	Parker et al. (2011a, b, c)

observed, but a detailed analysis of their structure was attempted, because the enormous work needed for this purpose did not seem to be justified at that time. In a later publication (Miller 1957), Miller confirmed formation of Gly, Ala, Asp, Glu, and Val and mentioned the detection of small amounts of eight prAAs. In addition to that, they identified succinic acid urea and *N*-methyl urea. In 1983, Schlesinger and Miller reported the detection of trace amounts of Ser and Val. Finally, two papers are worth mentioning in which after Miller's death analyses of ancient experiments were reported (Schlesinger and Miller 1983; Parker et al. 2011a, b). In those experiments, gas mixtures containing small amounts of HS were exposed to electric discharges to enable the formation of cysteine and methionine. Several sulfur containing amino acids were indeed detected, including Met, but cysteine or cystine was not found. These experiments allow for first important conclusion concerning the theory of CE, namely their results stand in sharp contrast to the reductionistic-deterministic concept (hypothesis IA in Sect. 1.2). Consider that formation of most biomonomers and biopolymers required for a successful CE is inscribed in the orbitals of atoms and molecules, one should expect that:

1. A significantly larger number of different prAAs should have been formed.
2. Non-protein AAS should only have been formed in traces or small amounts.
3. Functional by-products such as carboxylic acids that might hinder the formation of proteins should have almost absent.

The results discussed above and summarized in Table document a contrary trend, and this interpretation has been confirmed by the later experiments of the Miller group (Table 4.2) and by the numerous experiments contributed by other research groups (Table 4.3, 4.4, 4.5, 4.6, and 4.7).

Table 4.2 Syntheses of prAAS by means of electric discharges in a neutral atmosphere

Gas mixtures	prAAs	Non-prAAs	ω-AAs	*N*-Alkyl AAs	Refs.
CH_4, N_2, NH_3, H_2O	Ala, Gly, Asp, Glu, Ser, Thr, Leu, Ile, Val, Pro	α-Abu, α-Aibu, isoVal, norVal, isoLeu, norLeu, tert. Leu, isoSer, alloThr, α,ß-diaminopropionic a., α,γ-diaminobutyric a.	ß-Ala, ß-Abu, g-Abu, ß-Aibu,	Sar, *N*-ethyl Gly, *N*-propyl Gly, *N*-isopropyl Gly, *N*-methyl Ala, *N*-ethyl Ala, *N*-methyl-ß-Ala, Pipecol a.	Miller (1986)
Alkanes, ethylene, acetylene, H_2, H_2O	Ala, Gly, Asp, Glu, Ile, Leu, Pro, Ser, Thr, Val	No information	No information	No information	Ring and Miller (1984)
CO_2, N_2, H_2O	Aal, Gly, Asp, Glu, Ser	No information	No information	No information	Cleaves et al. (2008)

Table 4.3 Syntheses of α-amino acids by means of electric discharges in the gas phase

Starting materials	prAAs	Various AAs	Various reaction products	References
CO_2, NH_3, H_2O	Gly	No info	Formamide, oxalic acid amide	Löb (1913)
CH4, NH_3, H2O	Gly, Ala	α-Abu, ß-Ala, Sar	Substituted guanidines	Heyns et al. (1957)
CH_4, NH_3, H_2O, H_2S	Gly, Ala	α-Abu, ß-Ala, Sar	Ammonium rhodanide	Heyns et al. (1957)
CH_4, CO, NH_3, or CH_4, NH_3, H_2O	Gly, Ala, Asp, Glu, Lys	α-Abu, ß-Ala	No info	Pavlowskaya and Pasynski (1959)
CH_4, $C2H_6$, NH_3, H_2 H_2O	Gly, Ala, Asp, Asn, Ile, Pro	No info	Amines, amino acid amides, nucleobases	Oro (1963)
CH_4, NH_3, H_2O	Gly, Ala Asp, Glu, Ser, THr, Leu, Ile Lys	No info	No info	Grossenbacher and Knight (1959)
CH_4, NH_3, H_2O	Gly, Ala, Asp, Ser, His, lys, Ile	No info	No info	Matthews and Moser (1966)
CO2, N2, H2O	Gly, Ala	No info	No info	Plankensteiner et al. (2004)
CO_2, N_2, H2O	Gly, Ala, Val, Ser, Lys	No info	No info	Plankensteiner et al. (2006)
CO_2, N_2, H_2O	Gly, Ala, Val, Glu, Ser	ß-Ala, Gly-Gly (traces)	Orotic acid	Hirose et al. (1990)

Table 4.4 Syntheses of aminoacids in the gaseous phase by means of various energy sources

Starting materials	Energy source	prAAs	Byproducts	References
CH_4, NH_3, H_2O, CO_2, CO, H_2, N_2	X-rays	Various prAAs (not identif.)	Various amines	Dose and Rayowsky (1957)
CH_4, CO_2, N_2, NH_3, H_2O	X-rays	Gly, Ala, Asp.	Acetic a. succinic a., lactic a., pyruvic a., alkyl amines	Dose and Risi (1968)
CH_4, NH_3, H2O, H_2	UV	Gly, Ala, various prAAs	Various amines	Groth and Weysenhoff (1957)
CH_4, NH_3, H_2O	UV	Various prAAs	Aldehydes	Terenin (1959)
CH4, CO, NH3, H2O	UV	Various prAAs	norLeu, amines, hydrsazine, formaldehyde, urea	Dodonova and Siderova (1961)
CH4, NH3, H2O, H2S	UV	Gly, Ala, Asp, Glu, Ser, Cys	No info	Sacgan and Khare (1971)
CH4, NH3, H2O, H2	ß-Radiation	Various prAAs[a]	Urea, adenine and other by-products	Palm and Calvin (1962)

(continued)

Table 4.4 (continued)

Starting materials	Energy source	prAAs	Byproducts	References
CH4, NH3, H2O	ß-Radiation	Gly, Ala,	Glycine amide, nucleobases	Oro (1963)
CH4, NH3, H_2O, H_2S	ß-Radiation	No info	Taurine, cysteamine, cysteine acid	Choughuley and Lemmon (1966)
CO, NH3, H_2O	Protons (3 meV)	Gly, Ala, Asp, Ser	α-Abu, g-Abu, imidazole	Kobayashi et al. (1990)
CH_4, NH_3, H_2O	Heat: 950–1050 °C	Gly, Ala, Asp, Glu, Ser, Thr, Leu, Ile, Val, Pro, Phe, Tyr	ß-Ala, Sar	Harada and Fox (1964)
CH_4, NH_3, H_2	Heat: 950–1050 °C		α-Abu, allo-iso-Leu	Harada and Fox (1965)
CH_4, NH_3, H_2O, H_2	Heat:1300 °C	Gly, Ala, Asp, Glu, Ser, Thr, leu, Ile, Phe, Tyr	ß-Ala, allo-iso-Leu	Oro (1965)
CH_4, NH_3, H_2O	Heat:900–1150 °C	Gly, Ala, Asp. Glu, Leu, Val, Lys, Trp, Phe, Tyr	Acetaldehyde, propionaldehyde, acrolein, acetone, methanol, ethanol, var. dicarboxylic a, methyl and dimethyl amine, ethyl and diethylamine	Taube et al. (1967)
CO, NH_3, H_2	Heat up to 1000 °C	ß-Ala, Orn, Sar	Pyrimidine, purine	Yoshiro et al. (1971)
CH_4, NH_3, H_2O	Heat: 900–1060 °C	Gly, Ala	Succinic a.	Lawless and Boyden (1973)
CH_4, C_2H_6, NH_3, H_2O	Schock-wave	Gly, Ala, Leu, Val	No info	Bar-Nun et al. (1968), Bar-Nun (1975), Barak and Bar-Nun (1975)

[a]Various prAAs means that the individual AAs were not identified

Table 4.5 Syntheses of prAAs by means of UV radiation in solution or in the solid state

Starting materials	λ (nm)	prAAs	Various by-products	References
Glycol, glycerol, Na+, K+ and NH_{4+} nitrate + TiO_2	Sunlight	Various prAAs	No info	Dar and Mukerjee (1934)
CH_2O, KNO3, FeCl3	Sunlight	Gly, Ala	No info	Bahadur (1954)
CH_2O, N_2 + Mo oxides	Sunlight Gly, Ala	Gly, Ala	No info	Bahadur and Ranganayaki (1958)

(continued)

Table 4.5 (continued)

Starting materials	λ (nm)	prAAs	Various by-products	References
NH_4-HCO_3, KSCN, organic acids	283.7	Gly, Ala, Asp	No info	Deschreider (1958)
CH_2O, NH_4X + inorg. salts	Various λ	Gly, Ala, Glu, Ser, Ile, Val, Phe	Basic AAs	Pavlowskaya and Pasynski (1959)
CH_3OH, CH_2O, NH_3 + inorg. catalysts	184.8	Gly, Ala + various AAs	Formic a., hexamethylene tetramine	Reid (1959)
CH_3OH, $H_2O + NH_3$, or CH_3OH, H_2O + HCN	<180, T < 80 K	Gly, Ala, Ser	*N*-Formyl Gly, cycloserine, ethanol amine	Bernstein et al. (2002)
CH3OH, CO2, CO, H2O, NH3	<180 T < 80 K	Gly, Ala, Asp. Ser, Pro, Val	12 AAs, 2,5-diaminofuran, 2.5-diaminopyrrole, 1,2,3-triaminopropane	Munoz-Caro et al. (2002a, b)
CH_3OH, H_2O, NH_3	<180 T < 80 K	Gly, Ala, Asp, Ser, Pro, Hypro		Meierhenrich et al. (2002, 2005)

Table 4.6 Non-protein AAs that were identified after UV irradiation of simulated interstellar ice particles

Amino acids	Imino acids	Diamino acids
ß-alanine	Sarcosine	2,3-Diamino propionic a.
2-aminobutyric a.	*N*-Ethyl glycine	2,4-Diamino butyric a.
3-aminobutyric a.	*N*-Methyl alanine	3-Amino-2-(aminomethyl) propionic a.
4-aminobutyric a.	*N*-(aminomethyl) glycine	
3-amino isobutyric a.	*N*-(aminoethyl) glycine	
Aminomethyl butyric a.		
5-amino valeric a.		
Aminoethyl valeric a.		

After 1960, geologists and other scientists dedicated to research on CE became convinced that the strongly reducing atmosphere postulated by H. C. Urey did not exist at the time when the first biomolecules were synthesized, because almost all H2 had vanished to space. Furthermore, analyses of the oldest rocks and minerals (3.5–3.8 Billion years old) revealed the presence of oxidized iron. The majority concluded that the early atmosphere was strongly influenced by the exhalations of volcanoes so that ammonia was a relatively rare species and N_2 was predominant nitrogen compound in the atmosphere. Miller reacted immediately to this paradigm change and launched a new series of experiments using a neutral or only slightly reducing atmosphere (Table 4.2). However, these new experiments revealed that in the absence of reducing gases such as CH_4, NH_3, and H_2 no prAAs were formed with exception of Gly. This result means that when prAAs were formed in the early

Table 4.7 Syntheses of AAs in aqueous solution by means of various energy sources

Starting materials	Energy source	prAAs	By-products	References
NH3, Acetic a.	ß-Radiation	Gly, Asp	Diamino succinicv a.	Haselstrom et al. (1957)
CO2, NH3, H2O	γ-Radiation	Gly	Not identified AAS	Paschke et al. (1957)
NH3, Acetic a.	X-ray	Gly	ß-Ala	Dose and Ethre (1958)
Alkyl ammonium carbonates	X-ray	Gly, Ala, Asp, Lys	ß-Ala	Dose and Risi (1968)
NH_3, HCH	Heat (70 °C)	Various prAAs	No info	Oro and Kamat (1961)
NH_3, HCN, H_2O	Heat (90 °C)	Gly, Ala, leu, Ile, Val, Ser, Thr	α-Abu, ß-Ala, 2,3-diamino propionic a.	Lowe et al. (1963)
CO_2, NH_3 NH_2OH, HCN	Heat (70–100 °C)	Gly, Ala (Asp, Ser, Thr)?	ß-Ala	Oro et al. (1959)

stage of the CE, the atmosphere must have had a slightly reducing character, or local "hotspots" having a reducing atmosphere. At this point, it should be mentioned that such "hotspots" really exist for example in the form of "white smokers," which exhale CO and H_2. This is one reason, why certain biologist (e.g., Lane, Martin. Russel) consider white smokers to be the birthplaces of the CE (s. Sect. 2.6). Yet, this hypothesis has two serious drawbacks. Firstly, ammonia and HCN are missing, so that prAAs or nucleobases cannot be formed in these white smokers. Secondly, the alkaline character of the expelled water causes rapid racemization of prAAs and favors hydrolysis of peptide groups and of amide groups in the side chains of Asn and Gln.

The later "Miller experiments" (Table 4.2) yielded the already previously identified prAAs, but due to a more efficient analytical equipment, traces of Ile and Pro were also detected (Wolmann et al. 1972; Ring et al. 1972; Ring and Miller 1984; Miller 1986). In this way at least 9 out of the 20 prAAs needed for the construction of proteins were obtained. However, the number of non-protein amino and imino acids had also increased, namely up to 12 non-prAS, five ω-amino acids, and eight imino acids. As discussed in Sect. 9.4 in more detail, all these undesired and the CE unfavorable by-products were also extracted from meteorites. This conspicuous parallel suggests that the chemical course of the "Miller experiments" must be quite similar to that of the "space chemistry." A pertinent comparison was already discussed by Miller and coworkers in 1972 (Wolmann et al. 1972). Miller was also interested in the elucidation of the reaction mechanisms involved in the formation of the prAAs in his experiments. He considered the so-called Strecker synthesis of amino acids to be a suitable basis for his measurements and mechanistic hypotheses, in as much as he had found HCN and several aldehydes as by-products in his reaction mixtures (Miller 1957, 1959). However, in the classical "Strecker synthesis," the hydrolysis of the nitrile group

Scheme 4.1 Formation of α-amino and α-hydroxycarboxylic acids via a modified Strecker synthesis as proposed by S. L. Miller

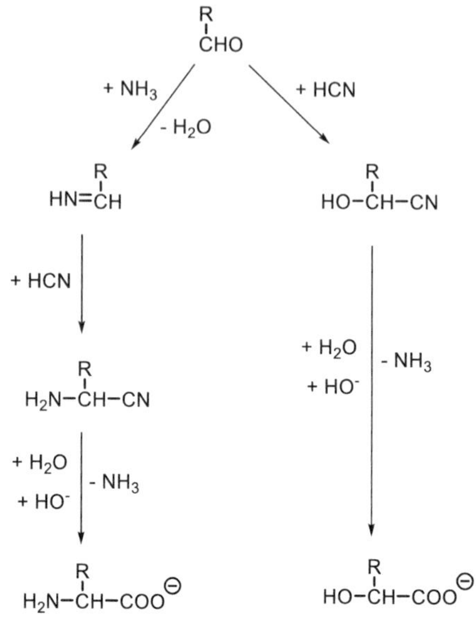

is achieved by heating moderately concentrated hydrochloric acid. Such a highly acidic reaction medium did certainly not exist on the early Earth, and even when, it had hindered any formation of proteins and polynucleotides. Miller being aware of this problem measured, how the concentration of ammonia, HCN, aldehydes, and prAAs varied in the course of his experiments and reached the following conclusion. He assumed that the hydrolysis of the nitrile group was catalyzed by OH^- ions and formulated the reaction pathway outlined in Scheme 4.1. In other words, Miller postulated that several intermediates, such as aldehydes or HCN, are formed in the gas phase, but that the final prAA synthesis occurs in the alkaline water phase, i.e., in the primordial soup. This reaction sequence might have taken place, but for the following reasons, it is far from a satisfactory explanation of the experimental results.

First, a significant concentration of ammonia in the primordial soup is a priori unlikely, because NH_3 is rapidly decomposed by UV light and because an efficient mechanism for its production does not exist. Furthermore, a significant concentration of ammonia would have prevented the evolution of proteins and RNAs. Second, numerous amino and imino acids were found in meteorites (s. Chap. 9) where a formation involving a hot ammonia solution is impossible, in as much as the temperatures in space are typically below −150 °C. Third, the similarity of the prAAs and of the non-prAAs found in Miller's experiments and in meteorites suggests that the same reaction sequence was operating in both cases. Yet, the space chemistry is mainly a chemistry based on reactions of radicals, and the electric discharges used by Miller also generate radicals. Therefore, the simultaneous formation of amino acids and hydroxy acids is more likely a result of radical reactions as illustrated in Scheme 4.2.

Scheme 4.2 Hypothetical formation of α-amino acids and α-hydroxy acids via radicals

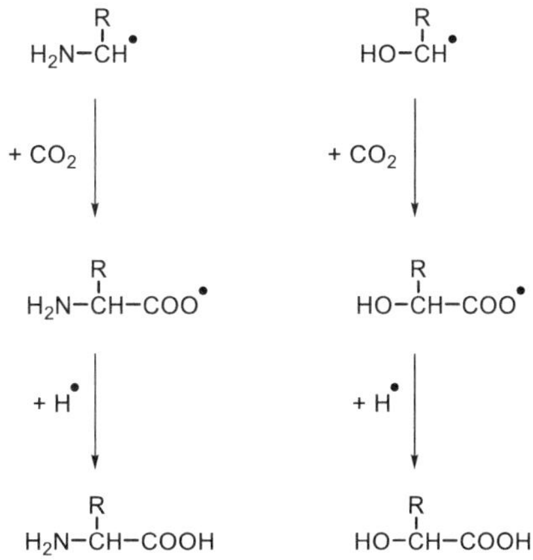

Finally, it should be emphasized that even when the results of all "Miller experiments" are summarized, the following prAAs were never obtained: Asn, Arg, Cys, Gln, Trp, Tyr.

4.2 Model Syntheses of Various Research Groups

Miller's publications in 1953 and 1955 stimulated numerous other researchers to perform similar experiments with broad variation of the starting materials, energy sources, and reaction conditions (Miller 1953, 1955) (Table 4.3, 4.4, 4.5, 4.6, and 4.7). Several research groups also studied the impact of electric discharges on gas mixtures; their results are summarized in Table 4.3. In the first seven experiments, reducing gas mixtures were used in analogy to the first series of "Miller experiments." Two new results were found. First, three research groups detected lys in their reaction mixtures (Pavlowskaya and Pasynski 1959; Grossenbacher and Knight 1959; Matthews and Moser 1966). In addition to Lys, those prAAs were detected that were previously reported by Miller. Second, two groups mentioned the detection of nucleobases as by-products, but these nucleobases were not unambiguously identified (Heyns et al. 1957; Oro 1963). This result is of particular importance, because it is not a convincing argument in favor of the CE hypothesis, when under the conditions "X," a couple of prAAs were synthesized but neither nucleobases nor nucleosides, whereas under the conditions "Y," two or three nucleobases were obtained, but no prAAs. In four publications, the formation of ß-Ala was mentioned and in three papers, the identification of α-Abu, but, unfortunately, most publications say only little or nothing about nature and quantity of by-products.

In numerous experiments, gas mixtures were exposed to various forms of energy (Table 4.4). In one series of experiments, the gases were irradiated with accelerated protons, and five prAAs, already reported by Miller, were obtained, but interestingly, imidazole was detected as by-product. This result is noteworthy, because imidazole plays an important role in model syntheses of oligo and polynucleotides as discussed in Chap. 7. Unfortunately, bombardment of the primordial soup with accelerated protons is not a trustworthy scenario for the beginning of the CE. Nine research groups used X-rays, UV, or ß-radiation as energy source, whereas seven groups used heating up to temperatures around 1300 °C as means to promote formation of prAAs. Remarkable is the detection of nucleobases in three out of 16 experimental set ups (Palm and Calvin 1962; Oro 1963; Yoshiro et al. 1971). Unfortunately, most authors did not explore to what extent non-prAAs were formed, and if so, ß-Ala and α-Abu were found in close analogy to "Miller's experiments." However, in contrast to Miller and coworkers, several research groups reported on the identification of basic by-products. In five cases, aliphatic diamines were detected. Noteworthy are also the experiments of Choughuley and Lemmon (1966), because those authors added H_2S to their gas mixture, but in agreement with the results of the Miller group (Parker et al. 2011a, b) neither cysteine nor cystine were found.

Further studies with UV light as energy source were conducted in such a way that the starting materials were dissolved in water or irradiated in the solid state (Table 4.5) in contrast to the experiments in Table 4.3 where gas mixtures were irradiated. In analogy to the "Miller experiments," Gly, Ala, Asp, Glu, and Ser were found. Yet, remarkable is the work of Munoz-Caro et al. (2002a, b), because those authors also identified 12 non-prAAs. Finally, the experiments of Meierhenrich and coworkers should be highlighted (Meinert et al. 2012). This group irradiated their substrates in so-called simulated interstellar ice particles. In addition to six prAAs, 18 (!) non-prAAs were detected, listed in Table 4.6. The significance of these experiments is discussed in the next section.

In the last Table of this chapter, those experiments are compiled, which are characterized by the treatment of substrates dissolved in aqueous solution with radioactive radiation or with heat. Again, the usual prAAs, already known from the "Miller experiments," were found. Yet remarkable are the experiments of Haselstrom et al. (1957) and Lowe et al. (1963), because diamino acids were detected. Certain α,ω-diamino acids play an important role in the RNA world hypothesis as discussed in Sect. 2.3. Also remarkable is the work of Oro et al. (1959), because formation of several aliphatic carboxylic acids is reported. Only in two more papers (out of 44 publications), formation of carboxylic acids is mentioned. However, after reading those publications, the author got the impression that most authors did not check the formation of these by-products (in contrast to Miller), because they ignored that carboxylic acid in large quantities might have had a decisive negative influence on the course of the CE.

Finally, the trustworthiness of the identification of prAAs needs a critical comment. Almost all analyses of the reaction mixtures were conducted by means of chromatographic methods. A commercial sample of the searched-after prAA was used for comparison to determine the correct R_f value. The reaction with ninhydrin is served for the optical identification. However, the vast majority of research groups did apparently not compare their reaction products with a broad variety of non-prAAs that might have been formed as by-products and might have by chance an identical R_f value. This point is important because the distances between the ninhydrin spots in a chromatogram of a complex reaction mixture may be short. A particularly problematic case is the apparent identification of Lys (Tables 4.3 and 4.4), which was not found by Miller. It is absolutely unlikely that radical reactions yielded Lys without Orn or any other α,ω-diamino acid. The space chemistry (Tables 4.6 and 9.4) demonstrates that several α,ω-diamino acids are formed together, and the relative abundance decreases with increasing number of atoms.

Another problem is the apparent identification of Trp, which was reported by only one author (Taube et al. 1967). It is again absolutely unlikely that in the high-temperature experiments performed by that author only one amino acid with *N*-heterocycle in the side chain was formed. Furthermore, the identification of Tyr, not observed by any other research group, is questionable. The prAAs formed at 1000 °C can only have survived by extremely rapid cooling (quenching) of the hot gases. Yet where did such reaction conditions exist on the early Earth? Volcanoes in close neighborhood of ice certainly present such a combination of largely different temperatures, but the exhalations of volcanoes are not known for the production of prAAs. Furthermore, the prAAs formed under such extreme conditions must have survived for millions of years until the RNA world had reached a state where L-prAAs were picked up from the primordial soup for protein syntheses. Finally, it should be emphasized that no research group has realized experiments yielding more than 12 different prAAs.

4.3 Discussion of Amino Acid Syntheses

Since the "Miller experiments" play a prominent role in the previous research activities and in the discussions concerning the hypothesis of CE, a detailed comment and evaluation of the prAA syntheses summarized in Tables 4.1, 4.2, 4.3, 4.4, 4.5, 4.6, and 4.7 should already be presented here in anticipation of Chap. 11. In an objective scientific evaluation of Miller's experiment, one has to differentiate between the historic merits of his work, on the one hand, and its value and importance as argument in favor of the CE. Concerning the historic importance, the rather skeptic biochemist R. Shapiro wrote (Shapiro 1987, p. 99):

> In the case of the Miller-urea experiment, the scientific community was impressed as well as the public. The work was cited repeatedly in the following years, incorporated into high school and university biology and geology texts, and featured in museum displays. It has become the classic best known experiment on the origin of life.

On the same page, he cites the chemist W. Day:

> It was an experiment which broke the logjam. The simplicity of the experiment, the high yields of the products and the specific biological products in limited number produced by the reaction was enough to show that the first step into the origin of life was not a chance event, but one that has been inevitable [....]. With the appropriate mixture of gases any energy source that can cleave the chemical bonds will initiate a reaction resulting in the formation of life's building blocks.

Afterwards the astronomer C. Sagan, who had dedicated much research activities to the exploration of life in the universe, was cited:

> The Miller-Urey experiment is now recognized as the single most significant step in convincing many scientists that life is likely to be abundant in the cosmos.

A high estimation of the Miller experiments in their historic role as door opener to the research on CE is, of course, still justified to day. Yet, what do these experiments say about the credibility of the CE in the first quarter of the twenty-first century? To answer this question correctly, one has to consider the problems that emerged from the experimental results and their reaction conditions.

1. The main reaction product of the Miller experiments was dark tar, and the amount steadily increases with increasing reaction time. The yields of all prAAs together did never exceed 5%.
2. At first, the low diversity of the prAAs isolated from those experiments. In the first full publication of Miller (1955), only two prAAs were isolated and three more prAAs were more or less clearly identified by chromatography. Taking into account, all the following experiments (Tables 4.1 and 4.2), a dozen of different prAAs were detected, but Asn, Gln, Arg, Cys, His, Tyr, and Trp were not found. With exception of Arg and Cys, these prAAs were observed by other research groups, but as discussed above their identification was not reliable, or the reaction conditions don't have any relevance for the origin of life on the early Earth.
3. Large amounts of non-prAAs and imino acids were obtained and in several experiments of Miller, sarcosine was even the main product. The ratio of non-prAAS (+imino acids)/prAAs is even more pronounced in the extracts of meteorites (Sect. 9.4) which demonstrates that the experiments of the Miller group were not exotic exception. The reactivity of most non-prAAs is as high as that of prAAs. Regardless of the polymerization mechanism, any polycondensation of prAAs and their derivatives had necessarily included incorporation of non-prAAs. However, even small amounts of Sar or ß-Ala in the polypeptide chain had the consequence that the characteristic secondary and tertiary structures were destabilized or distorted, with total loss of their biological function (Fig. 4.1). Have these numerous non-prAAs disappeared before the CE reached a stage where protein synthesis became feasible, how and why? Or, has the RNA world developed a selective protein synthesis in the presence of large amounts of non-prAAs? The author has only found two vague answers, one in the book of Wills and Bada (2000) and another one in the book of DeDuve (2002). The speculations of those authors will be discussed in Sect. 11.1. Yet, it should be emphasized at this point that no satisfactory hypothesis of a CE can be formulated without a convincing answer on the afore-mentioned questions.
4. Miller has found aliphatic carboxylic acid including hydroxy acids in molar ratios of 3/1 relative to the prAAs. In the extracts of meteorites, this ratio is even more extreme, so that the results of Miller rather underrate than overestimate the situation at the beginning of the CE. Most researchers who worked on model syntheses of prAAs did not care about this aspect. However, carboxylic acids

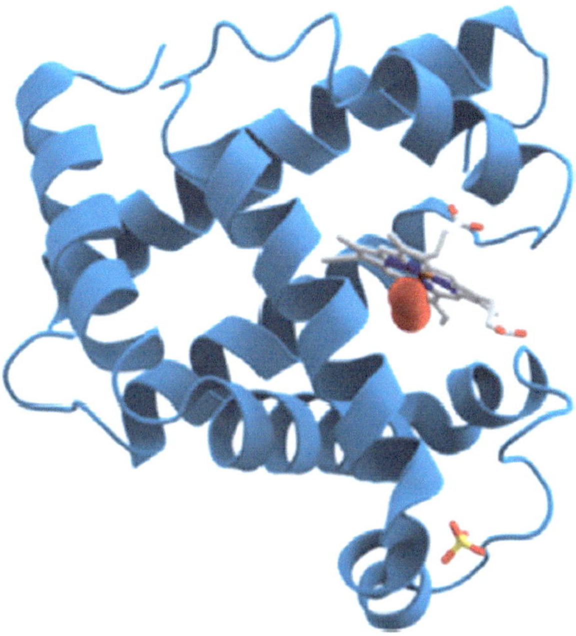

Fig. 4.1 Secondary and tertiary structure of myoglobin

react as chain stoppers in any polycondensation process yielding polypeptides regardless, if the carboxylic group is activated or the amino group. Eq. 3.8 in Sect. 3.1 and the calculated data presented in Table 3.1 clearly demonstrate that even at a molar ratio of 1/1 formation of peptide chains having a DP around 100 is impossible. A ratio of 3/1 hinders even a significant formation of a DP around 10. The results reported so far (Chap. 6) suggest that a prebiotic protein synthesis did not exist before the RNA world had developed a machinery capable of synthesizing useful proteins. Hence, it may be argued that the afore-mentioned negative influence of carboxylic acids was not effective. Yet, this hypothesis has to demonstrate that a prebiotic RNA machinery has emerged which worked successfully in the presence of formic acid, acetic acid, lactic acid, etc. The experimental evidence for the prebiotic existence of such a fantastic protein production does not exist, to the best knowledge of the author.

5. Regardless if the CO2H groups of the prAAs were formed by a kind of "Strecker Synthesis" (Scheme 4.1) or via reactions of radicals (Scheme 4.2) formaldehyde and higher aldehydes were necessary for syntheses of prAAs. The presence of larger amounts of formaldehyde is also necessary to explain formation of saccharides via the "Butlerow reaction," it is necessary to explain the existence of glycol aldehyde (needed for the formation of nucleotides, s. Chap. 8) and formaldehyde is a component of interstellar clouds. In other words, formaldehyde is one of the most fundamental molecules underlaying the CE. However, formaldehyde is a highly reactive electrophilic reagent which can easily react with any

kind of amino groups. Hence, significant amounts of formaldehyde and higher aldehydes can prevent the synthesis of proteins by blocking the amino end groups. Furthermore, it can react with the amino groups in the side chains of Lys, Orn, and Arg, and thus it can block the biological activity of proteins containing those prAAs. Moreover, formaldehyde can react with the amino groups or adenine (A), guanine (G), and cytosine (C) and prevent their participation in base pairing. Even, when only 5% of the A, G, and C are hydroxymethylated, the biological function of an RNA will be blocked. The question, who has taught formaldehyde to produce Gly, saccharides and nucleotides, but to be absent when syntheses of biologically active RNAs and proteins are needed is unanswered, and the author has not even seen any book or review article, where this question has been formulated.

Each one of the points 2–5 is per se sufficient to overthrow the hypothesis of CE, if a satisfactory explanation supported by experiments cannot be found. If one assumes that the vast majority of prAAs was delivered by comets, meteorites, and micrometeorites, the problems formulated in points 2 and 3 do not vanish. As pointed out in Chap. 9, more than 14 different prAAs were never extracted from meteorites and the problem of 3. becomes even more aggravated. The number and quantity of non-prAAs and imino acids that can prevent a successful synthesis of proteins are even much larger in the space chemistry. When the critique formulated at the end of Sect. 4.2, the results of other research groups compiled in Table 4.3, 4.4, 4.5, 4.6, and 4.7 and the points mentioned above are summarized, the only objective conclusion that may be drawn is that a prebiotic evolution of proteins was impossible on the basis of the results published so far. The historic merits of the "Miller experiments" and their usefulness as support of the CE hypothesis are two quite different aspects of his work. It is strange that even in books and review articles that appeared after 2000, the Miller experiments are exclusively presented in a positive light without any comment on the numerous problems and questions that arose from these experiments: Brack (1998), Fry (2002), Wills and Bada (2000), Zubay (2000), Day (2002), DeDuve (2002), Popa (2004), Thomas (2005), Luisi (2006), Rauchfuß and Mitchell (2008), Rehder (2010). A faint idea of a critique was given by Thomas (2005, translated by the author): "The critique of the "Miller experiments" is primarily directed against their low concentrations. It is unlikely that the amino acids formed in the primordial soup in low concentration will yield any other reaction product." However, this critique mainly concerns the primordial soup theory of Oparin and Haldane, but not the experiments of Miller.

At the end of this chapter, the discussion of point 2 should be broadened. Not more than 12 different prAAs were found in a single experiment and only 14 were detected in meteorites. This negative result along with the abundance of non-prAAs is, first of all, a disaster for the reductionistic-deterministic concept of the CE, but it is also a disaster for the "Protein-World" hypothesis (Sect. 2.1). Furthermore, it is a serious problem for the RNA world concept. Either, one has to assume that the lacking prAAs were formed under conditions favoring the synthesis of various RNAs, whereupon the variability of reaction conditions is strongly limited. Alternatively, one has to assume that the RNAs have developed the capability to synthesize the lacking prAAs themselves and to keep the unfavorable non-prAAs and other detrimental byproducts (e.g., carboxylic acids) away from the protein synthesizing machinery. Such an assumption involves a strong belief in chemical miracles and is far beyond the border of a reasonable scientific hypothesis.

References

Bahadur K (1954) Nature 175:1141
Bahadur K, Ranganayaki S (1958) Proc. Natl. Acad. Sci. (India) 27A:292
Barak J, Bar-Nun A (1975) Orig Life 6:483–506
Bar-Nun A (1975) Orig Life 6:109–1156
Bar-Nun A, Bar-Nun N, Bauer S, Sagan C (1968) Science 168:470–472
Bernstein M, Rworkin J, Sanford S, Cooper G, Allamandola J (2002) Nature 416:403–406
Brack A (ed) (1998) The molecular origin of life. Cambridge Univ. Press, Cambridge
Choughuley A, Lemmon R (1966) Nature 210:628
Cleaves H, Chalmers J, Lazcano A, Miller SL, Bada J (2008) Orig Life Evol Biosph 38:105–115
Dar N, Mukerjee S (1934) Nature 134:499
Day W (2002) How life began: the genesis of life on Earth. Foundation for new directions
DeDuve C (2002) Life evolving: molecules, mind and meaning. Oxford University Press, Oxford
Deschreider A (1958) Nature 182:228
Dodonova N, Siderova A (1961) Biofizika 6:199
Dose K, Ethre K (1958) Z Naturforsch 13b:784
Dose K, Rayowsky B (1957) Biochim Biophys Acta 25:225
Dose K, Risi S (1968) Z Naturforsch 23b:581
Fry I (2002) The emergence of life on Earth. Rutgers Univ. Press, New Brunswick
Grossenbacher K, Knight CA (1959) In: Fox S (ed) The origin of prebiotic systems. Academic Press, New York
Groth WV, Weysenhoff H (1957) Naturwissenschaften 44:511
Harada K, Fox S (1964) Nature 201:335
Harada K, Fox S (1965) In: Fox S (ed) The origin of prebiotic systems. Academic Press, New York
Haselstrom T, Henry C, Murr B (1957) Science 1925:350
Heyns K, Walter W, Meyer E (1957) Naturwissenschaften 44:385
Hirose Y, Ohmuro K, Sargoh M, Nayakama T, Yamagata Y (1990) Orig Life Evol Biosph 20:471–481
Kobayashi K, Tsuchiya M, Oshima T, Yamagaawa T (1990) Orig Life Evol Biosph 20:99
Lawless JG, Boyden CD (1973) Nature 199:219
Löb W (1913) Ber Dtsch Chem Ges 46:684
Lowe C, Rees M, Markham R (1963) Nature 199:219
Luisi P-L (2006) The emergence of life. Cambridge University Press, Cambridge
Matthews C, Moser R (1966) Proc Natl Acad Sci USA 56:1087
Meierhenrich U, Munoz-Caro G, Schutte W, Barbier B, Segovia A, Rosenbaum H, Thiemann W, Brack A (2002) ESA Spec Publ 518:25–30 (Proc. of 2. Eur. Workshop Exo Astrobiology)
Meierhenrich U, Munoz-Caro G, Schutte W, Thiemann W, Brabier B, Brack A (2005) Chem Eur J 11:4895–4900
Meinert C, Filippi J, DeMarcelius P, LeSergeant-d'Hendecourt L, Meierhenrihc U (2012) ChemPlusChem 77:186
Miller S (1953) Science 117:528
Miller S (1955) J Am Soc 77:2351
Miller S (1957) Biochem Biophys Acta 23:480
Miller S (1959) Science 130:245

Miller S (1986) Chim Scr 26B:5–11

Munoz-Caro G, Meierhenrich U, Schutte W, Barbier A, Arones-Segovia A, Rosenbaum H, Thiemenn W, Brack A, Greenberg J (2002b) Nature 416:403–406

Munoz-Caro GM, Meierhenrich U, Schutte W, Barbier B, Arones-Segovia A, Rosenbauer H, Thiemann W, Brack A, Greenberg J (2002a) Nature 417:403

Oro J (1963) Nature 197:862

Oro J (1965) In: Fox S (ed) The origin of prebiologic systems. Academic Press, New York

Oro J, Kamat SW (1961) Nature 190:442

Oro J, Kimball AP, Fritz R, Muster P (1959) Arch Biochim Biophys 85:115

Palm C, Calvin M (1962) J. Am. Soc. 84:2115

Parker E, Cleaves H, Callahab N, Dworkin J, Glavin D, Lazcano A, Bada J (2011c) OrigLife Evol Biosph 41:569–574

Parker E, Cleaves H, Callahan N, Dorkin J, Glavin D, Lazcano A, Bada J (2011b) Orig Life Evol Biosph 41:201–212

Parker E, Cleaves HJ, Dworkin J, Glavin D, Callahan N, Aubrey A, Lazcano A, Bada JL (2011a) Proc Natl Acad Sci USA 108:526

Paschke R, Chang R, Young D (1957) Science 125:881

Pavlowskaya T, Pasynski A, (1959) In: Clark E, Synge R (eds) The origin of life on Earth, Pergamon Press, New York

Plankensteiner K, Reiner H, Rode BM (2006) Mol Div 10:3–7

Plankensteiner K, Reiner H, Schranz B, Rode B (2004) Angew. Chem Int Ed 43:1886–1888

Popa R (2004) Between necessity and probability: searching for the definition and origin of life. Springer, Berlin

Rauchfuß H, Mitchell T (2008) Chemical evolution and the origin of life. Springer, Berlin

Rehder D (2010) Chemistry in space. Wiley VCH, Weinheim

Reid C (1959) In: Oparin AL et al (eds) The origin of life on Earth. Pergamon Press, London

Ring D, Wolmann Y, Friedman L, Miller SL (1972) Proc Natl Acad Sci USA 69:765

Ring D, Miller S.L. 8 (1984) Orig. Life 15:7–15

Sacgan C, Khare B (1971) Science 173:417

Schlesinger G, Miller L (1983) J Mol Evol 19:376–382

Shapiro R (1987) ORIGINS. Bentham and summit books: a division of Simon & Schuster, New York,

Taube M, Zdrojewski S, Samochoka K, Jezierska K (1967) Angew Chem 79:239

Terenin A (1959) In: Oparin A et al (eds) The origin of life on Earth. Pergammon Press, London, p 136

Thomas S (2005) Ursprung des Lebens. Fischer Taschenbuch, Frankfurt

Wills C, Bada J (2000) The spark of life. Perseus Publications, Cambridge

Wolman Y, Haverland WJ, Miller SL (1972) Proc. Acad. Sci. USA 69:809

Yoshiro D, Hyatsu R, Andersen B (1971) Geochim. Cosmochim. Acta 35, 927–938

Zubay G (2000) Origin of life on the earth and in the cosmos. Academic press, San Diego

5 Model Syntheses of Saccharides, Nucleobases, Nucleosides, and Nucleotides

We need to know more, we can know more and we will know more.

D. Hilbert

Nucleotides, the building blocks of RNA, consist of two types of organic compounds: D-ribose and nucleobases. The four nucleobases on which the RNA structure is grounded are the pyrimidine bases such as uracil and cytosine and the purine bases such as adenine and guanosine. Today, almost all scientists agree in that the DNA based on 2′-deoxyribose is a later invention of the evolution, following the earlier invention of RNA. The first primitive species of living organisms were capable of self-reproduction without containing DNA (such as the archaea still today). Hence, this chapter is focused on the discussion of those model experiments that contributed to a better understanding of the origin of RNA.

5.1 Syntheses of Saccharides

All speculations and all model syntheses concerning the prebiotic syntheses of saccharides deal with the s. c. "formose reaction." This cascade of addition and condensation reactions starts out from formaldehyde which can react with itself and includes aldol reaction, reversible aldol reaction, and aldehyde–ketone isomerization. This reaction cascade was first described in 1861 by A. Butlerov and therefore, this scenario was also frequently called "Butlerov reaction" or "Butlerov syntheses" (Scheme 5.1) (1861). Studies of the reaction mechanism began in the middle of the twentieth century with a major contribution of R. Breslow (1959). More recently, further detailed studies were published by Harsch et al. (1983a, b).

As demonstrated in Scheme 5.1, the first reaction product is glycolaldehyde which upon reaction with formaldehyde yields glycerol aldehyde and its isomer

H. R. Kricheldorf, *Life by Chemical Evolution?*
https://doi.org/10.1007/978-3-662-70752-4_5

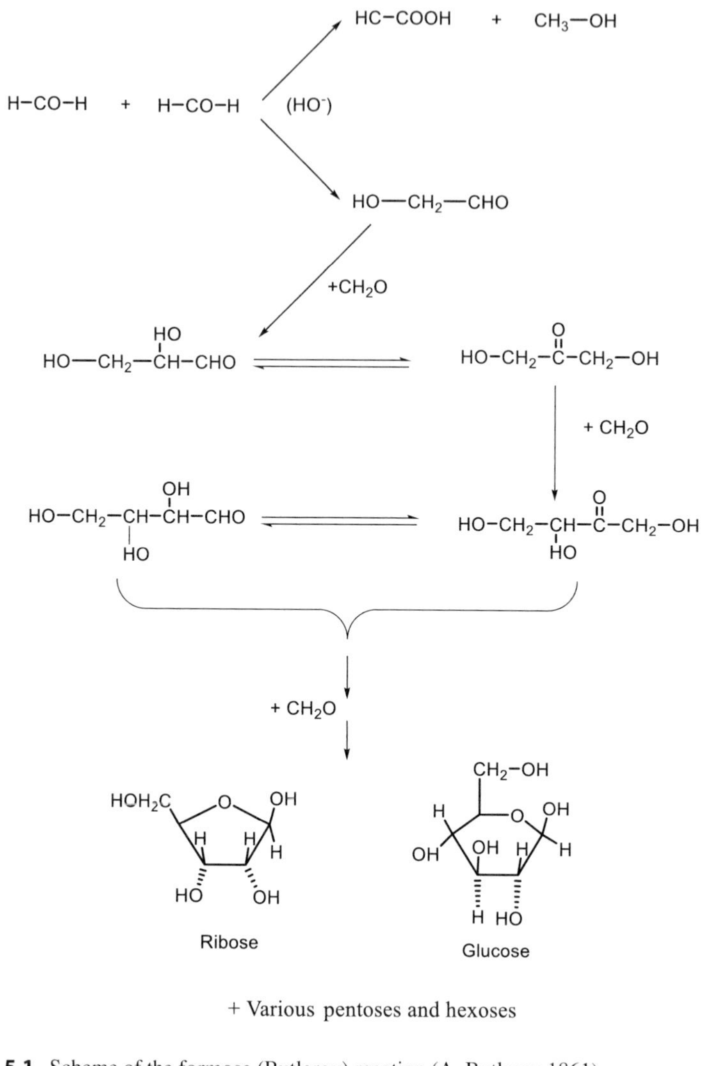

Scheme 5.1 Scheme of the formose (Butlerov) reaction (A. Butlerov 1861)

dihydroxyacetone. Addition of more CH_2O molecule yields the C4 sugars such as erythrose and threose. The next addition step yields pentoses including ribose, the key to the formation of nucleosides (for alternative syntheses of nucleosides, see Chap. 8). Formaldehyde and glycolaldehyde were detected in interstellar clouds, and they were extracted from meteorites together with traces of C_3–C_6 sugars. These findings let many scientists believe that formation reaction was the only source of ribose and C_6-saccharides, but the problem of stereoselective selection of D-sugars has not been solved yet.

Table 5.1 summarizes early studies dealing with the generation of formaldehyde and other aldehydes from CO_2. First studies concerning the synthesis of CH_2O from

Table 5.1 Syntheses of formaldehyde, glycolaldehyde, and saccharides

Starting material	Energy source	Aldehydes and saccharides	By-products	References
CO_2, H_2O KnO_3	UV	CH_2O + saccharides	Formaldehyde hydroxamic acid + var. by-prod.	Baly et al. (1922)
CO_2, H_2O	UV	CH_2O + Glyoxal	–	Groth and Suess (1938)
CO_2, H_2O, $FeSO_4$	He-ions 40 meV	CH_2O	Formic a.	Garrison et al. (1951)
CO_2, H_2O, kaolin	Heat	Trioses, tetroses, pentoses	–	Gabel and Ponamperuma (1967)
Glycolaldehyde, Glyceraldehyde	Heat + Zn-prolinate	Mainly tetroses Mainly pentoses	–	Kofoed et al. (2005)
CH_3OH, NH_3 H_2O	UV	CH_2O, glycolaldehyde, acrolein, and higher aldehydes	–	DeMarcellus et al. (2015)

CO_2 and H_2O were published by Baly et al. in 1922 and continued in 1938 by Groth and Suess (Baly et al. 1922; Groth and Suess 1938). Their experiments demonstrated that UV light with a wavelength of 180 nm and below is capable to cleave H_2O and to generate H radicals plus H_2. Other researchers used α-radiation (Garrison et al. 1951) or heat in the presence of kaolin (Gabel and Ponamperuma 1967) for similar experiments. In the classical Butlerov reaction, the aqueous formaldehyde is heated in the presence of strong bases. Such conditions might have existed inside "white smokers" (s. Chap. 2), but were extremely rare or absent on the early Earth. Therefore, several researchers studied reactions of CH_2O under weakly alkaline or neutral conditions, but again catalyzed by UV light (Oro and Cox 1962) or catalyzed by g-radiation (Ponamperuma 1965). Gabel and Ponamperuma (1967) heated aqueous formaldehyde in the presence of Al_2O_3, kaolin, or illite and found the typical reaction products of the formose reaction, but published their results without detailed identification of the reaction products. C. Ried and L. E. Orgel (1967) successfully used neat $CaCO_3$ or a combination of $CaCO_3$ and hydroxyapatite as v catalysts. However, those authors also observed that prolonged heating (up to 24 h) caused degradation of saccharides and draw the following conclusion:

> We do not believe that the formose reaction as we and others have carried it out, is a plausible model for the prebiotic accumulation of sugars. First, it requires concentrated solutions [of formaldehyde] ad second, the sugars formed decompose quickly. If formaldehyde is the prebiotic precursor of ribose, some methods of stabilization of the sugars is essential. The formation of ribosides of the natural bases – for example adenine - is one possibility, but attempts to bring about this condensation have not been successful. Instead rather unstable adducts were formed.

In the following years, numerous researchers concentrated a detailed analysis of the reaction products, above all on the identification of D-ribose, and only two reports in this area should be mentioned (Mizuno and Weiss 1974; Decker et al. 1982). Just on the limited basis of the analytical methods available at that time, Decker was able to conclude that the formose reaction yields more than 50 different products containing only traces of racemic ribose. G. Springsteen and G. Joyce wrote in 2004 the following comment (Springsteen and Joyce 2004):

> However attempts to devise a realistic prebiotic synthesis of nucleic acids from simple starting materials have been plagued by problems of poor chemical selectivity, lack of stereo- and regio-selectivity and similar rates of formation and degradation of some of the key intermediates. For example, ribose would have been only a small component of a higher complex mix of sugars resulting from the condensation of formaldehyde in a prebiotic world, In addition ribose is more reactive and degrades more rapidly compared with most other monosaccharides.

Even more negative is the comment of Robertson and Joyce (2012) in a review article:

> The classical prebiotic synthesis of sugars is the polymerization of formaldehyde (the formose reaction). It yields a very complex mixture of products including only a small proportion of ribose (Mizuno and Weiss 1974). This reaction does not provide a reasonable route

to ribonucleosides. However, a number of more recent experimental findings, to some extent, address this deficiency.

For example, Müller et al. reported in 1990 that the reaction of glycolaldehyde phosphate with a semi-equimolar amount of formaldehyde yields a complex mixture of tetrose, pentose, and hexose phosphates which contains ribose-2.4-phosphate as the main component (Müller et al. 1990). Such oligomerization reactions proceed even at temperatures as low as 20–25 °C if metal hydroxides having a layer structure (e.g., hydroxycalcite) are used as catalysts (Pitsch et al. 1995). In 2005, Kofoed et al. reported a higher yield of ribose, when glycerol was used as starting material, but those experiments do represent a realistic prebiotic scenario (Kofoed et al. 2005). Furthermore, the recent experiments of DeMarcellus et al. (2015) need mentioning. Those authors irradiated frozen solutions of CH_4, NH_3, and H_2O by means of UV light as simulation of the reaction conditions in interstellar clouds. As already found in other experiments simulating the interstellar chemistry (Sect. 9.3), a complex reaction mixture was obtained and not a specific formation of compounds favoring CE.

Finally, it is worth mentioning that two methods were found allowing for the stabilization of ribose. Sanchez and Orgel (1970) observed that a considerable stabilization of ribose occurs upon reaction with cyanamide, due to the formation of bicyclic 2-aminooxazolines (Scheme 5.2). An analogous cyclization also occurs, when other saccharides are reacted with cyanamide, but the 2-aminooxazoline of ribose has the tendency to crystallize spontaneously from the reaction mixture (Springsteen and Joyce 2004). This is, of course, a nice method for laboratory experiments, but its success is unlikely for a primordial soup containing a large excess of formaldehydes and other aldehydes which can also react with cyanamide. The assumption that compartments existed that selectively extracted pentoses and cyanamide is mere speculation. Furthermore, formation of cyclic esters of ribose and boric acid was assumed to have contributed to stabilization and enrichment of ribose (Ricardo et al. 2004). This hypothesis has three weak points. First, free boric acid is extremely rare on the Earth and metal borates don't react with ribose. Second,

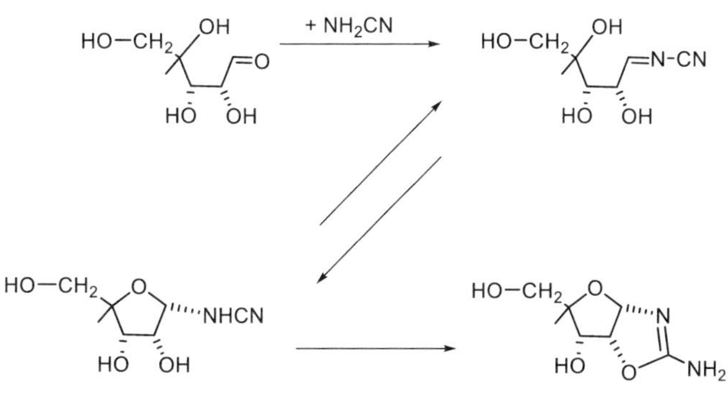

Scheme 5.2 Stabilization of D-ribose by rteaction witj cyanamide (Sanchez and Orgel 1970)

boric acid forms rather stable cycles with any kind of 1,2-diols having the OH-groups in cis-position. Third, a selective crystallization of the cyclic D-ribose borate does not take place. Furthermore, the hypothesis that cyclic derivatives of ribose might have played an important role for the formation of oligonucleotides, and thus, for the entire CE, has a serious short-coming. Conditions favoring formation of these cyclic derivatives are extremely unfavorable for any kind of polymerization process yielding oligo- or polynucleotides, because the OH groups in 2′- and 3'positon are blocked. Nonetheless, researchers who attribute an important role to the formation of cyclic borates of ribose are not rare. For example, an interesting review article of R. Scorei (2012) ends with the following statement:

> Because boric acid can stabilize ribose and form borate ester nucleosides, boron may have provided an essential contribution to the "pre-RNA World.

5.2 Model Syntheses of Nucleobases

In the years 1960 and 1966, the American chemist I. Oro published the first syntheses of a nucleobase under allegedly prebiotic conditions (structures and names of the nucleobases discussed in this work are summarized in Scheme 5.3). Oro obtained adenine under mild conditions by addition reactions of hydrogen cyanide in

Scheme 5.3 Structures of nucleobases and related heterocycles

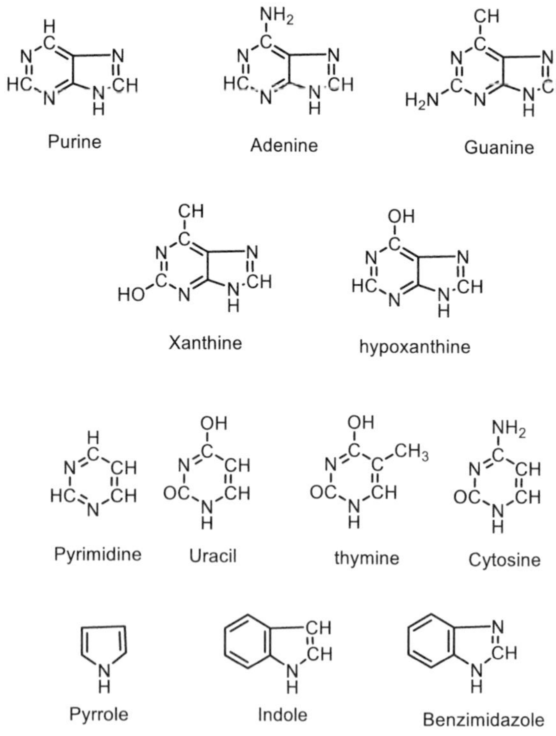

concentrated aqueous solutions (0.1 M—1.0 M) in the presence of small amounts of ammonia. Later more detailed studies revealed that aminomalononitrile and diaminomaleonitrile are formed as intermediates. A simplified version of the reaction pathway is outlined in Scheme 5.4 (Ferris 1968; Sanchez et al. 1967). Orgel and coworkers found that also 4-amino imidazole-5-carbonitrile and 4-amoino carboxamide were intermediately formed. Interestingly, these intermediates also play a role in the biosynthesis of purines in living organisms (Ferris and orgel 1966; Ferris 1968; Ferris et al. 1969). Sanchez et al. (1967, 1968) eventually detected small amounts of guanine in the reaction mixture, when stronger bases than ammonia were used as catalysts (Table 5.2). Furthermore, dicyano-diimidazole and methylated adenines were found when the oligomerization of HCN was performed in liquid ammonia (Wakamatsu et al. 1966; Yamada et al. 1968). Oligomerization of HCN in the presence of formaldehyde yields 8-hydroxymethyl adenine as the main reaction product (Schwarz and Baker 1989). A quite surprising result was published by Voet and Schwarz in 1983, because those authors reported that oligomerization of HCN followed by partial hydrolysis yielded uracil and orotic acid (Voet and Schwarz 1983). Yet, a confirmation of these results by other research groups does seemingly not exist.

What do these interesting results, which are of great importance from a historical point of view, say today about the probability that lining organisms emerged from CE? Four arguments vote against HCN as starting material for a prebiotic synthesis of purine bases. First, high concentrations are needed for a successful synthesis, and it is absolutely unlikely that such high concentrations existed in the primordial soup

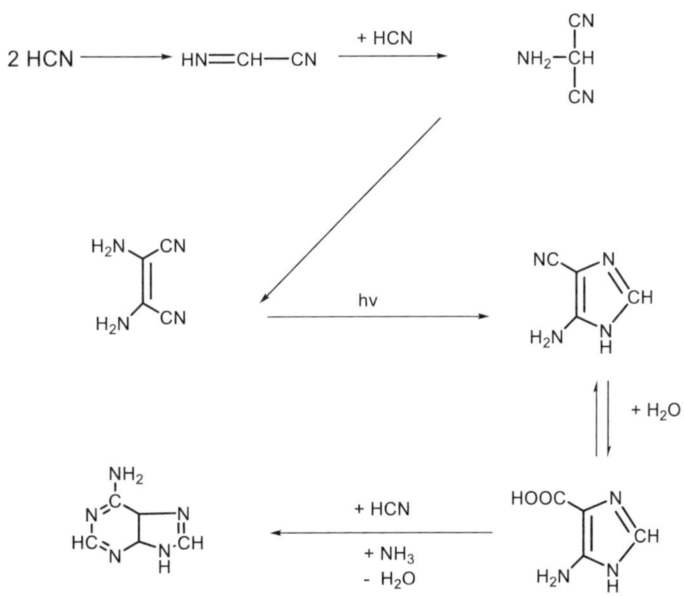

Scheme 5.4 Simplified reaction pathway of the synthesis of adenine from hydrogen cyanide

Table 5.2 Syntheses of nucleobases from HCN

Starting materials	Method	Nucleobases	By-products	References
HCN, NH_3	Heat (70 °C)	Adenine	Formamidine, oligom. HCN	Oro (1960); Oro & Kimball (1962)
HCN	Heat (90 °C)	Adenine	AAs + 75 diff. compounds	Lowe et al. (1963)
HCN	UV	Adenine, guanine	Urea	Ponamperuma (1965)
HCN + H_2O (pH 7–10)	Heat (up to 60 °C)	Adenine, Guanine	Malononitrile, hypoxanthine	Sanchez et al. (1967, 1968)
HCN in liqu. NH_3	Heat, pressure	Adenine	Dicyanoimidazole, Diaminomaleonitrile	Wakamatsu et al. (1966), Yamada et al. (1968)
HCN+ acetamidine in liqu. NH_3	Heat (120 °C), pressure	Adenine	Dicyanoimidazole, 2-methyl-, 8-methyl- or 2,8-dimethyl adenine	Yamada et al. (1968)
HCN	Oligomeriz. + Hydrolysis	Uracil	Orotic a.	Voet and Schwarz (1983)
HCN + CH_2O	–	Adenine (traces)	8-hydroxy methyl adenine, hydroxy acetonitrile	Schwarz and Baker (1989)

or in any compartment. Furthermore, volcanoes do not exhale significant amounts of HCN. Second, alkaline reaction conditions (pH > 8) are unlikely, except in special locations such as the "white smokers" (s. Chap. 2). However, the existence of "white smokers" on the early Earth is unproven, and the "white smokers" known today do not exhale significant amounts of HCN. Even more important is the argument that polycondensations of amino acids or nucleotides under alkaline conditions are impossible, and rapid racemization of prAAs under alkaline conditions is an additional counterargument (s. Sect. 10.3). Third, in the presence of formaldehyde, which is indispensable for the generation of saccharides or glycine, hydroxymethylated nucleobases will be formed in considerable amounts.

Several research groups selected formamide as starting material and substitute of HCN (Table 5.3). Since formamide may be understood as partially hydrolyzed HCN, it is not surprising that adenine was found as main reaction product in almost all those experiments. Only in the earliest study of this direction, Bredereck et al. (1962) unsubstituted purine was seemingly the sole nucleobase in the reaction mixture, but reliability and accuracy of the analytical method used at that time are questionable. The experiments performed after 2000 have in common that a mineral was added as catalyst (Saladino et al. 2001, 2003, 2004, 2007, 2008; Kumar et al. 2014). The most important and surprising result of those experiments is the finding that in addition to adenine, the pyrimidine bases cytosine and uracil or thymine were detected. This result means that the experiments of Saldino et al. and Kumar et al. came very close to the scenario required for a successful CE. However, in the experiments of Saladino et al. purine and various pyrimidine-type heterocycles were also formed, and it is not clear, if all N-heterocycles extant in the reaction mixture were

Table 5.3 Syntheses of nucleobases from formamide

Method	Nucleobases	By-products	References
Heat (mostly 100 °C)	–	Purine	Bredereck et al. (1962)
Heat + $POCl_3$ or organic acid chlorides	Adenine	No info.	Ochiai et al. (1968)
Heat + $CaCO_3$, SiO_2, kaolin, zeolite	Adenine, cytosine	Purine, 4-(3H)pyrimidine	Saladino et al. (2001)
Heat + montmorillonite	Adenine, uracil, cytosine	Purine, 5-hydroxy-xanthine, *N*-formyl purine	Saladino et al. (2004)
Heat + TiO2	Adenine, cytosine, thymine (traces)	Purine, 5-OH-methyl uracil, *N,N*-diformyladenine	Saladino et al. (2003)
Heat + FeS minerals	Adenine	Purine, 2-amiopurine Cyanamide, urea	Saladino et al. (2007, 2008)
Heat + Octacyano-molybdate salts	Adenine, cytosine, thymine	4(hydroxyacetyl)purine 4-(3-H)pyrimidinone	Kumar et al. (2014)
Heat + UV	Adenine, guanine	Hypoxanthine	Barks et al. (2010)
Mechanistic studies	Adenine	Purine	Hudson et al. (2012)

detected. All the heterocycles containing NH-groups may form nucleosides with ribose, and thus, may compete with the canonical nucleobases in the formation of mono-, oligo-, and polynucleotides. They play, so-to-say, the role of non-protein AASs in the peptide/protein chemistry (s. Chap. 4).

In this connection, it is of particular interest to mention the work of Senanayake and Idris (2006). Those authors irradiated formamide sprayed on the surface of TiO_2 crystals with UV light and reported that all the five nucleobases required for a successful CE were formed and that other N-heterocycles were absent. However, no compound was isolated, and the identification of the individual nucleobases was exclusively grounded on the molecule ion of the mass spectra. This method does, of course, not differentiate between isomers such as thymine and other methylated uracils. Furthermore, it is questionable if this analytical approach allowed for a reliable distinction between hypoxanthine and adenine or between guanine, xanthine, and 2,6-diaminopurine. It sounds miraculous that in such a simple experiment, all the nucleobases needed for a living organism were formed without any by-product, and the author of this book is, frankly speaking, skeptic, whenever chemical miracles exclusively yielding the desired products are reported. In contrast to the experiments catalyzed by minerals, experiments performed with heat or UV light, but without addition of catalysts yielded adenine, guanine, and hypoxanthine but not pyrimidine bases (Barks et al. 2010). In summary, the chemistry of formamide proved to be highly versatile and more promising for the hypothesis of CE than the chemistry of HCN. However, an important question, namely the question, if sufficiently high concentrations of formamide were available on the early Earth, remained unanswered. Nonetheless, three reaction pathways yielding formamide are imaginable: first, the hydrolysis of HCN; second, the reaction of NH_3 with CO (which requires a special catalyst); and third, a condensation of NH_3 with formic acid. These three reaction pathways are not alternatives and might have coexisted.

Starting out from liquid or solid materials, further model experiments were conducted with the goal to prepare pyrimidine bases (Table 5.4). A first successful synthesis of uracil was described in 1961 by S. W. Fox and K. Harada, who used malic acid and urea as reaction partners (Fox and Harada 1961). Urea was also used by Takenomoto and Yamamoto (1971), but maleic acid or fumaric acid served as reaction partners. Even heating urea with acetylene may yield small amounts of uracil (Subbaraman et al. 1980). All those experiments have in common that their authors did not care about the detection of other heterocycles, so that these reports do not carry much useful information. More successful and more informative were studies of Ferris et al. (1968, 1970, 1974), Robertson and Miller (1999), and Nelson et al. (2004). Those authors obtained cytosine in addition to uracil, when cyanoacetylene or cyanoacetaldehyde was used as starting materials. Finally, experiments of Nuevo et al. (2012) deserve to be mentioned. Those authors irradiated frozen solutions of pyrimidine with UV light to elucidate, if a transformation into uracil and/or cytosine can take place.

Following the footsteps of S. L. Miller, several research groups began after 1963 to synthesize nucleobases from mixtures of various gases and by means of various

Table 5.4 Syntheses of pyrimidine nucleobases from various starting materials

Starting materials	Method	Nucleobases	By-products	References
Malic a. + urea	Heat (130 °C)	Uracil	No info	Fox and Harada (1961)
Maleic a. or Fumaric a. + urea	Heat (160 °C)	Uracil	No info	Takenomoto and Yamamoto (1971)
Acetylene dicarboxylic a. + urea	Heat (100 °C) + Phosphoric a.	Uracil	No info	Subbaraman et al. (1980)
Cyanoacetylene + cyanate	30 °C in DMF or in water	Uracil, cytosine	Cyanovinyl urea	Ferris et al. (1968)
Cyanoacetaldehyde + guanidine	Heating in H_2O at pH >7	Uracil, cytosine	Diaminopyrimidine	Ferris et al. (1968)
Cyanoacetaldehyde + urea or Guanid.	Heat (100 °C)	Uracil, cytosine	Diaminopyrimidine	
Cyanoacetaldehyde + guanidine	100 °C or cold solution	Uracil, cytosine	Diaminopyrimidine	Robertson and Miller (1999)
Cyanaoacetaldehyde + urea	100 °C or lower temp.	Uracil, cytosine	No info	Nelson et al. (2004)
Pyrimidine in NH_3 or in NH_3 + H_2O	UV irrad. of frozen solution	Uracil, cytosine	Numerous not identif. products	Nuevo et al. (2012)

Table 5.5 Syntheses of nucleobases (mainly purines) from gaseous starting materials

Starting materials.	Method	Nucleobases	By-products	References
CH_4, NH_3, H_2O	ß-irradiation	Adenine	Formamidine polym.HCN	Ponamperuma et al. (1963a)
CH_4, H_2, NH_3, H_2O	4.5-MeV electrons	Adenine	Ribose, deoxyribose + var. saccharides	Ponamperuma (1965)
CO, H2, NH_3	Heat + pressure + Fe powder	Adenine, guanine, cytosine	Urea, biuret, cyanuric a., Melamine	Hayatsu et al. (1968)
CO, D_2, D_2O	Heat + pressure, Fe/Ni on Al_2O_3	Adenine, guanine, uracil, thymine	Methyl urea, hydantoin, xanthine, melamine, methyl guanidine, 5,6-dimethyl uracil	Hayatsu et al. (1972)
CO, H_2, NH_3	Heat + pressure, magnetite	Adenine, guanine, uracil, thymidine, cytosine	Urea, biuret, xanthine, melamine, ameline + 11 AAs	Anders et al. (1974)
CH_4, NH_3, H_2O	Electric discharges	Adenine, guanine	Isocytosine, 4-aminoimidazole-5-carboxamide	Yuase et al. (1984)
CO, N_2, H_2O	Irradiation with protons	Uracil	Imidazole	Kobayashi and Tsuji (1997)
CO, N_2, H_2O	Irradiation with protons	Adenine, guanine, uracil	5-Hydrouracil, 4,5-dihydrouracil, orotic a., nicotinic a.	Myakawa et al. (2002)
CO, N_2, H_2O	Plasma quenching	Uracil, cytosine	Amino acids (no details)	Myakawa et al. (1999)
CO, N_2, H_2O	Plasma quenching	No info	No info	Myakawa et al. (2002)

sources of energy (Table 5.5). In the experiments of Ponamperuma et al. (1963a), mixtures of CH_4, NH_3, and H_2O were exposed to ß-radiation. Interestingly, adenine was the only nucleobase detected in the reaction mixture, but formamidine and oligomerized HCN were also found. Those results suggest that HCN played the key role in the complex reaction process. In later experiments, Ponamperuma (1965) added a small amount of H_2 and used 4.5 MeV electrons for the irradiation. Again adenine was the only nucleobase in the reaction mixture, what underlines its exceptional stability. The same author also reported that ribose, 2′-deoxyribose and other saccharides were also formed in his experiments. This result is of great importance, because it demonstrated for the first time the simultaneous synthesis of a nucleobase and pentoses, but unfortunately those results were never confirmed by another research group, and it is questionable if the reaction conditions were suited for the formation of nucleosides and nucleotides.

In 1968, Hayatsu et al. exposed a mixture of CO, NH_3, and H_2O or D_2O to heat and pressure in the presence of iron isolated from meteorites and obtained cytosine, adenine, and guanine together with other nitrogen compounds (Hayatsu et al. 1968). Those authors concluded:

> The experiments were intended to simulate condensation in the solar nebulae.

Yet, only traces of cytosine and uracil were detected in meteorites (s. Chap. 9.), and it is questionable to what extent the Hayatsu experiments simulate the chemistry in space, in as much as the concentration of iron in interstellar clouds is too low to enable a role as catalyst. In two later papers (Hayatsu et al. 1972; Anders et al. 1974), the experimental conditions were varied, above all the catalysts, but those authors claimed again that their experiments match the reaction conditions in space. Those experiments are particularly important and noteworthy, because they yielded all the five nucleobases needed for the synthesis of RNA and DNA. Furthermore, apparently amino acids were also found albeit not characterized in detail. Hence those experiments come close to the reaction conditions needed for an initiation of the CE. However, those experiments need a critical comment. First, reports of other research groups describing a reproduction, further elaboration and optimization of those experiments are missing. Second, as clearly discussed in connection with the "Miller experiments" (Sect. 4.1), the presence of H_2 in the atmosphere of the early Earth is unlikely. Third in addition to the five canonical nucleobases, numerous other N-heterocycles were found in the reaction mixture, namely 1,3,5,6-dimethyluracil, xanthine, ameline, melamine, and cyanuric acid. All these heterocycles can form nucleosides with ribose and compete with the five canonical nucleosides for the formation of nucleotides and finally play the role of comonomers which prevent formation of the typical RNA sequences. Nobody has explained how and why these N-heterocycles have disappeared prior to the start of the CE.

In 1984, Yuase et al. reported that treatment of a CH_4, NH_3, and H_2O mixture with electrical discharges yielded adenine, guanine, and a little cytosine but no uracil (Yuase et al. 1984). When the same gas mixture was irradiated with accelerated protons, uracil was detected in the reaction mixture by two groups (Kobayashi and Tsuji 1997; Myakawa et al. 2002), but surprisingly only the second group mentioned formation of adenine and guanine. Nonetheless, the work of Kobayashi and Tsuji is remarkable, because it is the only report which mentioned small amounts of imidazole as by-product. Possibly, other researchers did not care about the identification of this heterocycles. Yet, the rare detection of imidazole in model experiments parallels the observation that only traces of imidazole were detected in meteorites, if any.

This aspect is of great interest, because imidazole played a key role in numerous experiments of research groups which studied syntheses of polypeptides or oligo- and polynucleotides (s. Chaps. 6 and 7). The research of Kobayashi (Myakawa et al. 1999, 2000) also explains the influence of plasma quenching on mixtures of CO_2, N_2, and H_2O. Surprisingly either pyrimidine bases or purine bases were exclusively formed, depending on the reaction conditions.

When all these model experiments are evaluated with respect to their usefulness as pillars of the CE hypothesis, the following conclusions may be drawn. Irradiation with accelerated electrons or protons and plasma quenching are all nice academic methods, but they have certainly not played a decisive role for a CE on the early Earth. Gas mixtures containing H_2 are not relevant for reasons discussed in Chap. 1. Hence, it may be said that among the experiments listed in Table 5.5 only those of Yuase et al. represent appositive contribution to the hypothesis of CE. Concerning the experiments of Table 5.4, it should be mentioned that maleic acid cyano-acetonitrile and cyanoacetaldehyde react with amino groups upon heating, and thus, would have prevented an evolution of peptides and proteins. A weak point of all experiments of Table 5.5 is also the fact that their authors did not care about formation of prAAs under the given reaction conditions. A successful CE requires simultaneous formation of prAAs and nucleotides, and it is rather a day dream than a reasonable hypothesis to assume that two sorts of compartments existed, one sort exclusively producing nucleobases and nucleotides and another one synthesizing prAAs.

5.3 Syntheses of Nucleosides

When compared to the numerous experiments dealing with syntheses of prAAs and when compared with the number of experiments dedicated to syntheses of nucleobases, only few studies of model experiments concerning the formation of nucleosides under prebiotic conditions were published. This matter of facts does, of course, not result from a low interest in this issue, it results from greater difficulties to find suitable reaction conditions and it results from less successful experiments. The first experiments designed to simulate a prebiotic formation of nucleosides were published by G. Schramm in 1961 and 1962. (Table 5.6). Upon heating of ribose and adenine with a reaction product of diethyl ether and polyphosphoric acid, he obtained small amounts of adenosine. Ponamperuma's group (Ponamperuma et al. 1963b; Ponamperuma and Kirk 1964) also reported that small amounts of adenosine and 2′-deoxy adenosine were detectable, when mixtures of ribose and adenine or 2′-deoxyribose and adenine were exposed to UV light in the presence of polyphosphoric acid. However, Reid et al. (1967) reported a few years later that those experiments were not reproducible. More successful was heating of ribose and purine bases with mixtures of phosphate and $MgCl_2$, but the yields were below 10%. And the amino groups of adenine and guanine were involved in side reactions (Fuller et al. 1972a, b). The same authors also found low yields of purine nucleosides, when the purine bases and ribose were heated in sea water until complete evaporation was achieved. Pyrimidine bases did not react under these conditions (Fuller et al. 1972b). L. E. Orgel (2004) wrote a negative comment on the prebiotic formation of nucleosides in a review article of 2004. However, more recently the research group of M. W. Powner and J. D. Sutherland found reaction pathways for syntheses of nucleoside 2′,3′-cyclophosphates that don't need free ribose as starting material (s. Sect. 5.4, Chap. 7).

Table 5.6 Syntheses of nucleosides

Starting materials	Methods	Nucleosides	By-products	References
Adenine + ribose or deoxyribose	Heat + ethyl metaphosphate	Adenosine	No info	Schramm et al. (1962), Schramm (1965)
Adenine + ribose	UV + H_3PO_4	Adenosine (0.01%)!	No info	Ponamperuma and Mariner (1963)
Adenine + deoxyribose	UV + H_3PO_4 or HCN	Deoxy adenosine (up to 7%	No info	Ponamperuma and Kirk (1964)
Cyanamide, cyanoacetylene, ribose, arabinose	Heat (100 °C) + UV	ß-cytidine 5% after photoisomer	α-cytidine	Sanchez and Orgel (1970)
Adenine, guanine, xanthine, hypoxanthine + ribose	Heat + H_3PO_4 + $MgCl_2$	Adenosine, guanosine, inosine	Ribosylation of amino groups	Fuller et al. (1972a)
Adenine, guanine	Heat: Evaporation of seawater solutions	Adenosine, guanosine, inosine, xanthosine	Pyrimidines did not react	Fuller et al. (1972b)
N-formyl aminopyrimidine + ribose	Heat (65–100 °C), alkaline pH	Adenosine	α-Anomers of ribose	Becker et al. (2016)

5.4 Syntheses of Nucleotides

The first syntheses of nucleotides (for example, of structure and terminology, s. Scheme 5.5) were described by several research groups in the years 1963–1965 (Ponamperuma 1965; Steinman et al. 1964; Neumann and Neumann 1964). Solutions of adenosine in liquid phosphoric acid were irradiated with UV light, and a mixture of adenosine and its mono-, di-, and triphosphate was obtained. However, those authors did not check to what extent the other OH-groups were phosphorylated. Mixtures of adenosine and its 5′-phosphates were also obtained by Waehneldt and Fox (1967) upon heating without UV radiation at temperatures below 160 °C. The chromatographic analyses revealed phosphorylation in 2′- and 3′pos-tion, and a significant selectivity for the 5′-position was not observed. Several research groups studied the evaporation of aqueous solutions of adenosine and phosphoric acid at 160 °C simulating the drying of the primordial soup on the hot slopes of volcanos (Ponamperuma and Mark 1965; Schwarz and Ponamperuma 1968; Schwarz 1969, 1974). However, this procedure is rather non-realistic considering the low thermal and hydrolytic stability of RNA and considering the racemization of prAAs (s. Chaps. 6, 9).

The research group of Orgel worked with adenosine at 100 °C and studied the usefulness of different phosphorylating agents (Lohrmann and Orgel 1968, 1971;

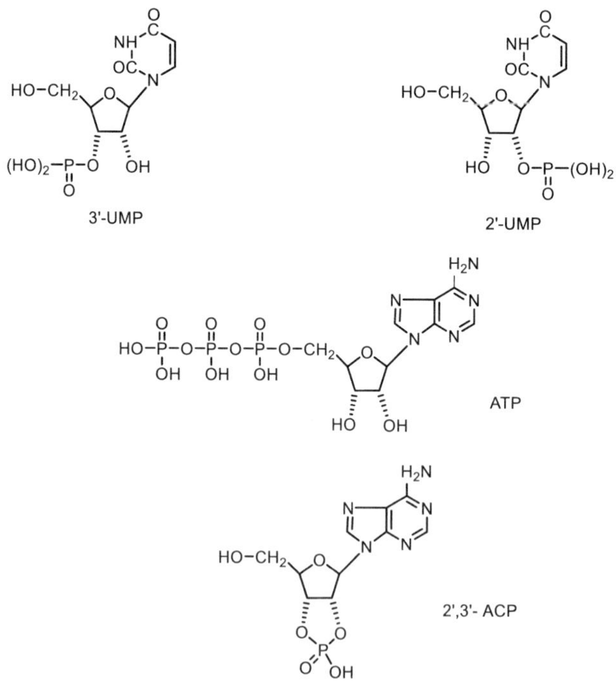

Scheme 5.5 Various kinds of nucleotides and their acronyms

Handschuh et al. 1973; Etaix and Orgel 1978). Particularly interesting was here the observation that pyrophosphate groups were formed, when metal hydrogen phosphates were heated with urea or cyanates. With addition of urea, it was also feasible to activate hydroxyapatite as phosphorylating agent (Lohrmann and Orgel 1971; Handschuh et al. 1973; Reimann and Zubay 1995). In this connection, it should be mentioned that pyrophosphates are also formed by heating of orthophosphates with cyanoacetylene (Ferris et al. 1970). Those authors (Ferris et al. 1984) also studied the usefulness of diamino succinic dinitrile as dehydrating agent for phosphorylation reactions. Several research groups used trimeta-phosphates as phosphorylating agents and obtained mixtures of nucleosides and their di- and triphosphates (Yamagata et al. 1995; Cheng et al. 2002).

All experiments dealing with phosphorylation of nucleosides have in common that the yields of phosphorylated products were below 50% and in most cases below 10%. Therefore, several authors conducted their experiments in organic solvents and higher yields were indeed obtained (Phillip and Seliger 1977; Schofstall and Koko 1978; Schofstall et al. 1979). However, such reaction conditions are far from realistic prebiotic conditions. An interesting result of several research groups is the finding that 2ß′,3′-cyclic phosphates of the ribonucleosides were formed (Table 5.7). These nucleosides derivatives are, of course, not suited for polycondensation, but they are useful for ring-opening polymerization, when preformed RNA sequences are present that serve as templates (s. Chap. 6).

Table 5.7 Syntheses of nucleotides starting out from nucleobases or from nucleosides

Starting materials	Energy source	Products	References
Adenosine + liquid H_3PO_4	UV + Water	Adenosine mono-, di- and triphosphates	Ponamperuma (1965), Ponamperuma and Mariner (1963), Ponamperuma et al. (1963a, 1963b)
Adenosine, guanosine, cytídine, uridine + var. Phosph.	Heat (160 °C) + Water	2′, 3′and 5′-Monophosphates of all four nucleosides	Ponamperuma and Mark (1965)
Adenosine and Graham salt	Heat (100 °C) + Water	2′-,3′and 5′monophosphates of adenosine	Schwarz and Ponamperuma (1968)
Adenosine, Deoxyadenosine	Heat, Na-trimetaphosphate	Mainly 3′-phosphates	Schwarz (1969)
Uridine, ammonium oxalate	Heat (90 °C) + Apatite	Mainly 5′-monophosphate	Schwarz (1974)
Adenosine, guanosine, cytosine, uridine, thymidine	22 °C, polyphosphoric a.	Mono-, di-, and triphosphates of all nucleosides	Waehneldt and Fox (1967)

(continued)

Table 5.7 (continued)

Starting materials	Energy source	Products	References
Uridine, cyanamide KOCN, NC.CO-NH_2	Heat (65 °C), H_2O with pH 6–8	Mainly 2′, 3′-cyclophosphate	Lohrmann and Orgel (1968)
Uridine, Urea, NH_4Cl	Heat (65–100 °C), hydroxyapatite	2′-,3′-, and 5′-monophosphates and 2′,3′-Cyclophosphate	Lohrmann and Orgel (1971), Handschuh et al. (1973)
Various nucleosides	Na-trimetaphosphate	3′- and 5′-diphosphates	Etaix and Orgel (1978)
Various deoxynucleosides, formamide	H_2O with pH >7	3′- and 5′-triphosphates	Phillip and Seliger (1977)
Adenosine	Na-trimetaphosphate	Mainly 2′,3′-cyclophosphate	Schofstall and Koko (1978), Schofstall et al. (1979)
Adenosine	H_2O, 37 °C, dihydrogen phosphate	Mainly 5′-monophosphate	Yamagata et al. (1995)
Adenosine 5′-monophosphate, KOCN	H_2O	Adenosine 5′-di- and triphosphate	Yamagata (1999)
Various nucleosides, urea	Heat (90 °C), H_2O, Hydroxyapatite	Mainly 5′-phosphates	Reimann and Zubay (1995)
Adenosine	Heat (60 °C), metal ions, Na-trimetaphosphate	5′-triphosphate (ATP) + 2′,3′Cyclophosphate	Cheng et al. (2002)
Glycolaldehyde, glyceric aldehyde, cyanamide, cyanoacetylene	22 °C, H_3PO_4	Cytidine 2′,3′-cyclophosphate	Powner et al. (2009)

A completely new approach to the synthesis of nucleosides and nucleotides, which does not rely on the availability of free ribose, was elaborated by Powner et al. (2011). Those authors started out from four chemicals that were presumably present on the early Earth, in as much as, two of them, glycolaldehyde and cyanoacetylene were detected in interstellar clouds. The main reaction product was the cyclic 2′,3′-phosphate of cytidine, and it was found that under irradiation with UV light, a selective of the amino group is feasible, so that the cyclic phosphate of uridine was also obtained.

This new approach is the additional advantage that the problem of regioselective glycosidation of the nucleobases is solved. However, a new problem came into sight, namely the stereochemistry of C1 of the ribose. The reaction sequence elaborated by Powner et al. (outlined in Scheme 5.6) yields at first the α-anomer. Irradiation with UV light allowed for partial transformation into the desired ß-anomer, but the yields were extremely low (<5%). Therefore, those authors

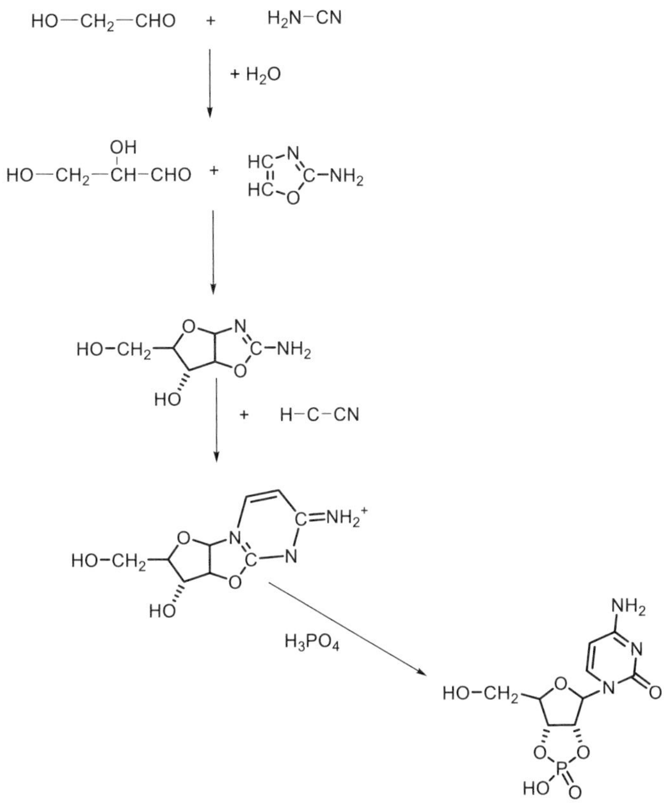

Scheme 5.6 Direct synthesis of cytidine from glycolaldehyde and cyanamide

elaborated an alternative reaction pathway characterized by the intermediate formation of a 2-thiouracil derivative (Scheme 5.7). That reaction product gave a significantly higher yield of the ß-anomer upon irradiation with UV light. Finally, the cyclic 2′,3′-phosphates of cytidine and uracil were isolated (Xu et al. 2017). This synthesis was hailed by those authors as particularly important and attractive, because 2-thiopyrimidine nucleosides also exist in transfer RNAs of living organisms. Two more weak points of the original Powner-Sutherland approach were also intensively studied. The first problem is the fact that stepwise synthesis which requires isolation and purification of reaction intermediates does not meet the requirements of a prebiotic reaction pathway. The second problem is the need of a similar approach to the synthesis of purine nucleotides.

To tackle that problem, Powner and coworkers developed "multicomponent one-pot procedures" which worked in aqueous solution (Powner et al. 2010). 2-Aminooxazole and 2-aminoimidazole served as starting materials, two compounds which result from reactions of HCN and cyanamide. Several interesting ribose derivatives were isolated that might be useful as intermediates in syntheses of both pyrimidine and purine nucleotides, but in 2021, the final goal was not achieved

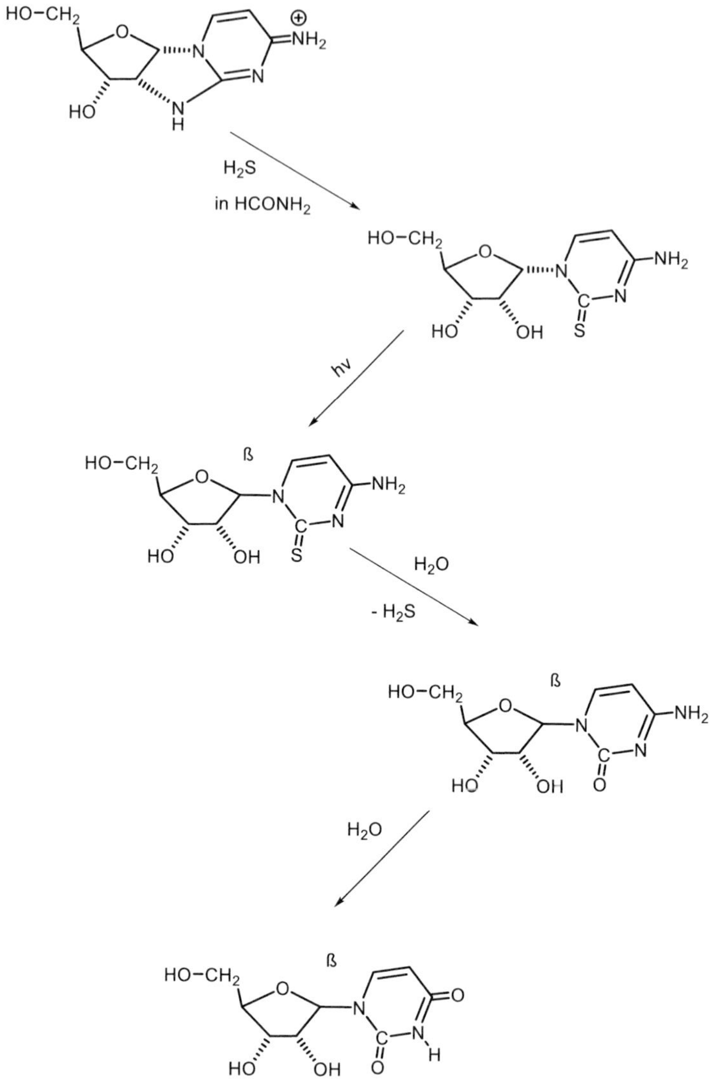

Scheme 5.7 Efficient synthesis of cytidine and uridine via a thiouracil intermediate

yet. More successful were several attempts to synthesize purine ribonucleotides in a more direct way (Stairs et al. 2017). As outlined in the simplified Scheme 5.8, those experiments were at least successful with regard to the synthesis of 8-oxopurine nucleotides. Finally, it should be mentioned that a promising approach to direct syntheses of 2′-deoxy nucleotides was elaborated, meaning reaction pathways avoiding an intermediate isolation of 2′-deoxyribose. Cyanamide and 2-mercaptoacetaldehyde proved to be useful starting materials. Despite significant

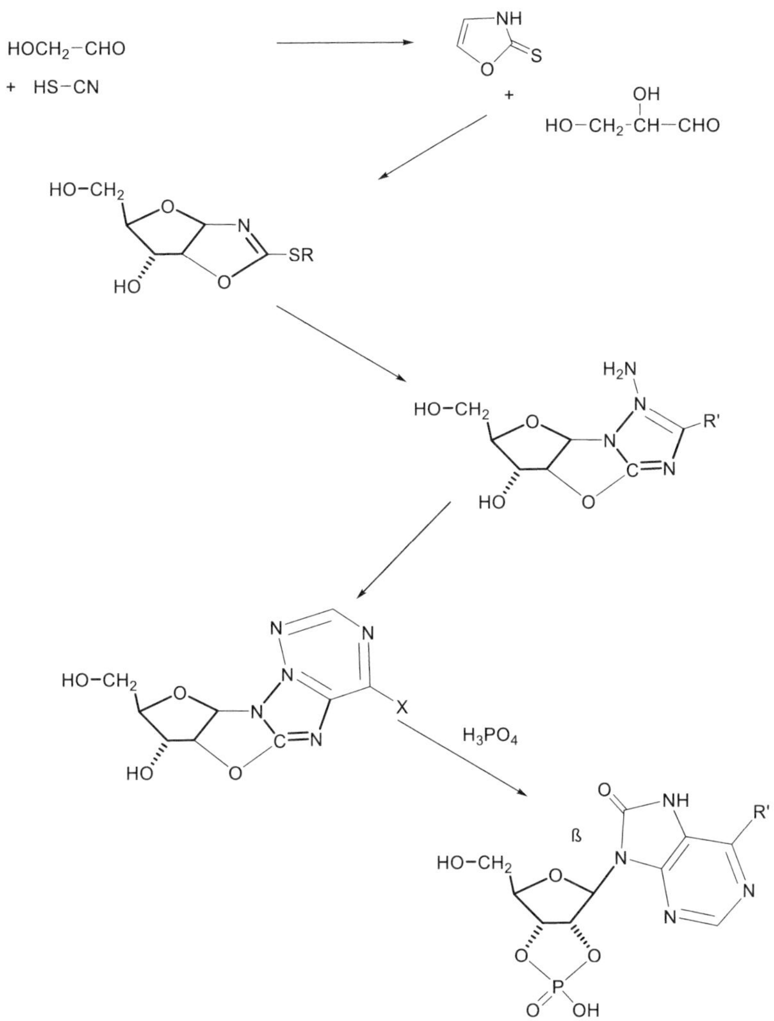

Scheme 5.8 Direct synthesis of ß-(8-oxopurine) ribonucleotides from glycolaldehyde and thiocyanic acid without free ribose as an intermediate

progress, the final goal, isolation of ß-2′- deoxy-ribonucleotides was not successful yet in the year 2017.

Further, more recent studies of nucleotide syntheses under plausible prebiotic conditions starting out from glycolaldehyde and other ribose-free starting materials published by the Powner group and by the Sutherland group should here be mentioned outside Table 5.7: Xu et al. (2020, 2021), Green et al. (2021), Coldville and Powner (2021), Whitaker and Powner (2022), Liu and Wang (2022), Feldmann et al. (2023), Bonfio et al. (2020), Liu et al. (2020a, 2020b, 2021).

5.5 Final Evaluation of the Various Model Syntheses

First of all, it should be emphasized that the Powner-Sutherland syntheses of nucleotides represent an important progress and support for the hypothesis of the "RNA-World" concept.

However, the following arguments need to be considered, if the usefulness of the afore-mentioned model syntheses for the hypothesis of CE should be evaluated. At first, it is necessary to take into account that the Powner-Sutherland experiments require numerous highly reactive electrophilic starting materials, such as formaldehyde, glycolaldehyde, glycerol aldehyde, mercaptoacetaldehyde hydrogen cyanide, and cyanoacetylene. All these electrophilic compounds can easily react with aliphatic amino groups and, thus, prevent formation of oligo- and polypeptides when significant concentrations are present in the prebiotic soup or in compartments/vesicles. Since the desired reaction steps are not quantitative, by-products are formed and may pile up and finally hinder a continuation of the CE. For example, nucleosides and nucleotides of non-canonic nucleobases such as xanthine or hypoxanthine may be formed and result in useless RNA sequences. Finally, it must be taken into account that both nucleosides and nucleotides are missing in meteorites, although the starting materials of the Powner-Sutherland syntheses belong to the "space chemicals." This findings raise doubts the Powner-Sutherland syntheses played a key role in the prebiotic CE (Powner et al. 2012).

References

Anders E, Hayatsu R, Studier M (1974) Orig Life 5:57
Baly E, Heilbron I, Hudson D (1922) Chem Soc 121:1078
Barks H, Buckley R, Grieves G, DiMauro E, Hud N, Oriando T (2010) ChemBioChem 11:1240
Becker S, Thomas I, Deutsch A, Gehrke T, Mayer P, Zipse H, Carell T (2016) Science 362:833
Bonfio C, Russel D, Green N, Mariani A, Sutherland J (2020) RSC Chem Sci 11:10688–10097
Bredereck H, Effenberger F, Rainer G, Schlosser H (1962) Liebigs Ann Chem 659:133
Breslow R (1959) Tetrahedron Lett 1:22
Butlerov A (1861) Liebigs Ann Chem 120:296
Cheng C, Fan C, Wan R, Tong C, Miao Z, Chen J, Zhao Y (2002) Orig Life Evol Biosph 32:219
Coldville W, Powner M (2021) Angew Chem Int Ed 60:10526–10530
Decker P, Schweer H, Pohlmann R (1982) Chromatographia 244:281
DeMarcellus P, Meinert C, Mygorodska T, Nahon I, Buse T, Le Sergeant d´Hendrecourt L, Meierhenrich UJ (2015) Proc. Natl. Acad. Sci. USA 112:965–970
Etaix E, Orgel L (1978) Carbohydr Nucleosides Nucleotides 5:91
Feldmann J, Skaanning M, Lommel M, Kernmayr T, Mayer P, Carell T (2023) ChemsitryEurope. https://doi.org/10.1002/ceur202300013
Ferris J (1968) Science 161:53
Ferris J, Goldstein G, Beaubeu D (1970) J Am Chem Soc 92:6598
Ferris J, Kuder J, Catalano A (1969) Science 166:756
Ferris J, Orgel L (1966) J Am Chem Soc 88:1074
Ferris J, Sanchez R, Orgel L (1968) J Mol Biol 33:693
Ferris J, Yamagawa H, Dudgeon P, Hagan W, Mallare T (1984) Orig Life Evol Biosph 15:29
Ferris J, Zamek O, Altbuch A, Freiman H (1974) J Mol Evol 3:101
Fox S, Harada K (1961) Science 133:1923

Fuller W, Sanchez R, Orgel L (1972a) J Mol Biol 67:25
Fuller W, Sanchez R, Orgel L (1972b) J Mol Evol 1:249
Gabel N, Ponamperuma C (1967) Nature 216:451
Garrison W, Morrison D, Hamilton J, Benson A, Calvin M (1951) Science 114:416
Green N, Xu J, Sutherland J (2021) J Am Chem Soc 143:7219–7236
Groth W, Suess H (1938) Naturwissenschaften 26:77
Handschuh G, Lohrmann R, Orgel L (1973) J Mol Evol 2:251
Harsch G, Harsch M, Bauer H, Voelter W (1983b) Z Naturforsch 38B:1269
Harsch G, Harsch M, Voelter W (1983a) Z Naturforsch 38B:1257
Hayatsu R, Studier M, Matsuola S, Anders E (1972) Geochim Geophys Acta 36:555
Hayatsu R, Studier M, Oda A, Fuse K, Anders E (1968) Geochim Geophys Acta 32:176
Hudson J, Eberle J, Vachhani R, Rogers L, Wade J, Krishnamurthi R, Springsteen G (2012) Angew Chem IntEd 51:34
Kobayashi K, Tsuji Z (1997) Chem Lett 26:903–904
Kofoed J, Reynolds J, Darber T (2005) Org Biomol Chem 3:1850
Kumar A, Sharma R, Kamaluddin MR (2014) Astrobiology 9:769
Liu Y, Wang Y (2022) Nat Chem 14:719–727
Liu Z, Wo L-F, Bond S, Sutherland J (2020a) RSC Chem Comm 56:13565
Liu Z, Wo L-F, Bonfio C, Ruissel D, Sutherland J (2020b) Nat Chem 12:1025–1028
Liu Z, Wu L-F, Kufner C, Sasselov D, Fischer W, Sutherland J (2021) Nat Chem 13:1126–1132
Lohrmann R, Orgel L (1968) Science 161:64
Lohrmann R, Orgel L (1971) Science 171:400
Lowe C, Rees H, Markham R (1963) Nature 199:219
Mizuno T, Weiss A (1974) Adv Carbohyd Chem Biochem 29:175
Müller D, Pitsch S, Kittaka K, Wagner F, Winter C, Eschenmoser A (1990) Helv Chim Acta 73:1410
Myakawa S, Murayawa K, Kobayashi K, Sawanka A (1999) J Am Chem Soc 121:8144
Myakawa S, Murayawa K, Kobayashi K, Sawanka A (2000) Orig Life Evol Biosph 30:557
Myakawa S, Yamanashi H, Kobayashi K, Cleaves H, Miller S (2002) Proc Natl Acad Sci USA 99:14628
Nelson K, Robertson M, Levy M, Miller S (2004) Orig Life Evol Biosph 31:221
Neumann W, Neumann M (1964) Univ. Rochester AEC reports, UR 656, Rochester, NY
Nuevo M, Milan S, Sandford S (2012) Astrobiology 12:295
Ochiai M, Manumoto S, Shimazu H, Morita K (1968) Tetrahedron 24:5731
Orgel L (2004) Crit Rev Biochem Mol Biol 39:99
Oro J (1960) Biochem Biophys Res Commun 2:407
Oro J (1961) Nature 191:1193
Oro J, Cox A (1962) Fed Proc 21:80
Oro J, Kimball P (1962) Arch Biochem Biophys 96:293
Phillip M, Seliger H (1977) Naturwissenschaften 64:273
Pitsch S, Eschenmoser A, Gedulin B, Hui S, Arrhenius G (1995) Orig Life Evol Biosph 25:297
Ponamperuma C (1965) In: Fox S (ed) The origin of prebiological systems. Academic Press, New York
Ponamperuma C, Kirk P (1964) Nature 203:400
Ponamperuma C, Lemmon R, Mariner R, Calvin M (1963a) Proc Natl Acad Sci USA 49:732
Ponamperuma C, Mariner R (1963) Nature 198:1199
Ponamperuma C, Mark R (1965) Science 148:1221
Ponamperuma C, Sagan C, Mariner R (1963b) Nature 199:222
Powner M, Gerland B, Sutherland J (2009) Nature 23:400
Powner M, Sutherland J, Szostak J (2010) J Am Chem Soc 132:16677
Powner M, Zheng S, Szostak J (2012) J Am Chem Soc 134:13889
Powner M, Sutherland JD, Szostak JW (2011) Synlet. 14:1956–1964
Reid C, Orgel L (1967) Nature 216:455
Reid C, Orgel L, Ponamperuma C (1967) Nature 216:936
Reimann R, Zubay G (1995) Orig Life Evol Biosph 29:299

Ricardo A, Carrigan M, Olcott A, Brenner S (2004) Science 303:196
Robertson M, Joyce G (2012) Cold Spring Harb Symp Quant Biol 4(5):a003608
Robertson M, Miller S (1999) Nature 375:772
Saladino R, Ciuambecchini U, Crestini C, Costanzo R, Negri R, Di Mauro E (2003) Chembiochem 4:514
Saladino R, Crestini C, Ciambecchini U, Cicirello F, Costanzo G, DiMauro E (2004) Chembiochem 5:1558
Saladino R, Crestini C, Cicirello F, Costanzo G, DiMauro E (2007) Chem Biodivers 4:694
Saladino R, Crestini C, Costanzo G, Negri R, DiMauro E (2001) Bioorg Med Chem 9:1249
Saladino R, Nen V, Crestini C, Costanzo G, Graciotto M, DiMauro E (2008) J Am Chem Soc 130:15512
Sanchez R, Ferris J, Orgel L (1967) J Mol Biol 30:223
Sanchez R, Ferris J, Orgel L (1968) J Mol Biol 38:121
Sanchez R, Orgel L (1970) J Mol Biol 47:531
Schofstall A, Barto R, Ramos D (1979) Orig Life 12:143
Schofstall A, Koko B, (1978) In: Noda H (ed) Orig Life, Proc. 2nd ISSOL meeting, p 193
Schramm G (1965) In: Fox S (ed) The origin of prebiological systems. Academic Press, New York
Schramm G, Grötsch H, Pollmann W (1962) Angew Chem Int Ed 1:1
Schwarz A (1969) Chem Commun 1969:1395
Schwarz A (1974) Biochim Biophys Acta 281:477
Schwarz A, Baker C (1989) Science 345:1102–1104
Schwarz A, Ponamperuma C (1968) Nature 218:441
Scorei R (2012) Orig Life Evol Biosph 42:5
Senanayake S, Idris H (2006) Proc Natl Acad Sci USA 103:1194
Springsteen G, Joyce G (2004) J Am Chem Soc 126:9578
Stairs S, Nikma A, Bucar D, Zheng S, Szostak J, Powner M (2017) Nat Commun. https://doi.org/10.1038/nccomms.15278
Steinman G, Lehmann R, Calvin M (1964) Proc Natl Acad Sci USA 57:27
Subbaraman A, Kazi Z, Choughuley A, Chada M (1980) Orig Life 10:343
Takenomoto K, Yamamoto I (1971) Synthesis 3:154
Voet S, Schwarz A (1983) Orig Life 12:84
Waehneldt T, Fox S (1967) Biochim Biophys Acta 134:1
Wakamatsu H, Yamada Y, Saito T, Kumashiro I, Takenishi T (1966) J Org Chem 31:2035
Whitaker D, Powner W (2022) Nat Chem 14:766–774
Xu J, Chmela V, Green N, Russel D, Janicki M, Gora R, Szabla R, Bond A, Sutherland J (2020) Nature 582:60–66
Xu J, Grenn N, Russell D, Liu Z, Sutherland J (2021) J Am Chem Soc 143:14482–14486
Xu J, Tsanakopoulos M, Magnani C, Szabla R, Sponer J, Gora R, Sutherland J (2017) Nat Chem 9:303
Yamada Y, Kumashiro I, Takenishi T (1968) J Org Chem 33:642
Yamagata Y (1999) Orig Life Evol Biosph 29:511
Yamagata Y, Inoue H, Inomata K (1995) Orig Life Evol Biosph 25:47
Yuase S, Flory D, Basil B, Oro J (1984) J Mol Biol 21:76

Model Syntheses of Oligo- and Polypeptides

6

Experience is the fruit of revised past errors, hence, we have to err from time to time.

I. Newton

6.1 Thermal Polycondensations

First experiments dealing with polycondensations of prAAs or their alkyl esters were performed before world war I and were not related to the hypothesis of CE. For comparison with later experiments (sd. Table 6.1), the work of Balbiano (1900, 1901) and Curtius and Benrath (1904) should be mentioned. Those authors heated prAAs to temperatures above 100 °C hoping that condensation reactions with elimination and evaporation of water occur. However, in those and later studies of Katchalski (1951) and Meggy (1954), the authors observed that heating of dry prAAs results mainly in decomposition and not in polycondensation. Therefore, glycerol was added as reaction medium for glycine and then formation of oligoglycines was detected. In similar experiments, Fox et al. found (Fox and Middlebrook 1954; Fox et al. 1957) that in addition to oligopeptides, cyclic dimers (2.5-diketopiperazines) were formed.

After these non-satisfactory experiments, S. W. Fox designed a new approach, which reduced the extent of decomposition by using a large excess of Glu (in a few cases Asp) or Lys (Fox and Harada 1960; Fox et al. 1962; Harada and Fox 1958). Large excess means a molar fraction around 60%. Even without addition of a liquid reaction medium, a kind of copolypeptides was obtained. At first, mixtures of two prAAs with an excess of Glu were used and finally mixtures of all 20 prAAs. The polycondensation products were called proteinoids. Temperature and time were varied, but preferentially a combination of 120 °C for 1 week or 170–180 °C for 6–10 h were applied. The analytical methods of that time did not allow from a detailed characterization of these proteinoids, but the Fox group was able to demonstrate that

H. R. Kricheldorf, *Life by Chemical Evolution?*
https://doi.org/10.1007/978-3-662-70752-4_6

Table 6.1 Syntheses of oligopeptides and proteinoids by thermal condensation of prAAs

Monomers	Methods	Reaction products	References
Gly	Heating in glycerol (150–170 °C)	OligoGly + 2,5-Diketopiperazine	Balbiano (1900, 1901)
Gly + Ethyl hippurate	Heat (190 °C)	OligoGly	Curtius and Benrath (1904)
Gly	Heat (+H_2O)	OligoGly	Meggy (1954)
Gly, Asp.Leu, Val, Phe	Heat (200 °C)	Oligomers + 2,5-diketopiperazines	Fox and Middlebrook (1954), Fox et al. (1957)
Pyroglutamic a, Ala, Asp, Gly,Leu, Ser, Val	Heat (196 °C or 199 °C)	Poorly characterized oligopeptides	Harada and Fox (1958)
Gly + Glu (excess)	Heat (175–190 °C)	Copolypeptides (Mn up to 12 kDa)	Harada and Fox (1958)
All 20 prAAs but excess Aps, Glu	Heat (175–180 °C)	Proteinoids (Mn = 3.6–8.0 kDa)	Fox and Harada (1960), Fox et al. (1962)
18 prAAs without Cys and Hypro	Heat (170 °C) + excess Asp, Glu	Proteinoids	Krampitz (1962)
Various prAAs	Heat + activated Al_2O_3	Mainly dipeptides and 2,5-diketopiperazines	Bujdak and Rode (2001, 2002, 2003)

the sequences were not random in nature. After fractionation by dialysis, molecular weights (number average Mn) in the range of 6000–10,000 were obtained, so that the term copolypeptides or proteinoids were indeed justified. Evidence for a chemical structure mainly based on peptide groups was achieved by hydrolytic degradation with proteolytic enzymes, such as pepsin or chymotrypsin (Fox and Harada 1960). However, this analytical method did not prove that amide groups based on g-carboxylic groups of Glu or ε-amino groups of Lys were absent. In the case of proteinoids rich in Lys-units, evidence for branching and crosslinking via ε-amino groups was found (Harada 1959; Harada and Fox 1965a; Heinrich et al. 1969). The analytical methods of that time did not allow the researchers to elucidate to what extent thermally unstable prAAs, such as Tre or Hypro had suffered thermal degradation.

A particular important problem of the Fox–Harada experiments was the question, if and to what extent the thermal polycondensations were accompanied by racemization. For two reasons, Fox and Harada were not able to determine a degree of racemization for the complete proteinoids. First, the complexity of their composition (incl. partial degradation during polycondensation) makes it difficult to synthesize a standard. Second, those authors used mixtures of L- and D,L-prAAs to lower the costs of their experiments. However, when L-Asp or L-Glu were used in large excess, it was possible to determine at least roughly the optical purity of these components. The result was a disaster, because complete racemization was found for Asp and partial, but considerable racemization for Glu (Fox and Dose 1977). These results are in good agreement with later studies of other research groups concerning the sensitivity of various prAAs to racemization (Chap. 10). For two reasons, only racemic prAAs may have existed at the beginning of the CE. The first reason were the high temperatures which accelerate racemization of prAAs even in neutral water (Chap. 10). Second, any synthesis of prAAs from achiral starting materials has necessarily resulted in formation of racemic prAAs. Every hypothesis of a CE has, therefore, to explain, how homochiral proteins emerged from a mixture of racemic prAAs in the primordial soup or in compartments. The Fox–Harada experiments demonstrate the opposite trend, more or less racemized proteinoids stating out from L-prAAs. A convincing solution of this problem has never been presented so far.

Immediately after 1960, Fox and Harada caused quite a stir and attracted interest and appreciation of numerous research groups, in as much as several proteinoids showed (weak) catalytic activities (s. Table 6.12 in Fox and Dose 1977). Furthermore, Fox and coworkers reported that proteinoids are form microspheres which under certain conditions were capable to undergo a kind of "cell division." Therefore, Fox was convinced that the proteinoids played a decisive role in the early stage of CE. The enthusiasm about the results of Fox and coworkers had the consequence that Fox was appointed director of new institute of molecular and cellular evolution at the university of Miami. However, after 1980, critique of Fox's overinterpretation of his proteinoid chemistry became stronger and stronger, for example, because weak catalytic activities, such as those of the proteinoids were also detected in various oligopeptides and the tendency of "cell division" was not confirmed by other researchers. The problem of racemization was not even recognized at that time, and in the year 2022, one may conclude that the Fox–Harada experiments have certainly historic merits by stimulating research on CE, but they have not made a significant contribution to a plausible and convincing theory of CE. A comment of C. Wills and J. Bada (Wills and Bada 2000, p. 5) is worth to be cited in this connection:

> Fox's experiment was an intriguing one, but, unfortunately, he stepped beyond the boundaries of his experimental data. He claimed that as his amino acids were heated and joined together, they tended to order themselves into reproducible sequences rather than simply linking up at random. He characterized this process as the very beginning of life, constituting a "biomacromolecular big-bang."

A further severe point of critique which concerns all model syntheses of oligo- and polypeptides will be discussed at the end of this chapter.

For the sake of completeness, it should be mentioned more recently that Bujdak and Rode (2001, 2002, 2003) studied again thermal polycondensations of prAAs in the temperature range of 85–200 °C. The amino acids were deposed on the surface of activated Al_2O_3 prior to heating, but without excess of Asp, Glu or Lys again only mixtures of oligopeptides, 2,5-diketopiperazines and degradation products of prAAs were obtained.

6.2 Polycondensation of Amino Acid-Phosphoric Acid Anhydrides

Almost at the same time as Fox and Harada were active, E. Katchalsky began with polycondensation experiments based on prAAs activated by means of phosphoric acid anhydrides. After the assassination of Katchalsky by a terrorist, those experiments were continued and expanded by his former coworker M. Paecht-Horrowitz (Table 6.2). At first, Gly and H-Gly-Gly-OH were polycondensed with trimetaphosphate, but only moderate yields of low oligomers were obtained (Katchalski and Paecht 1954). When using the same method a few years later, Chung et al. (1971) were not more successful despite broad variation of the reaction conditions. Shortly before, Schramm and Wissmann (1958) studied the polycondensation of h-D,L-Ala-Gly-Gly-OH with phosphorous pentoxide. A tripeptide was used as "monomer" to avoid the rapid formation of a 2,5-diketopiperazine competes with chain growth. With this approach, oligopeptides having DPs up to 24 were identified by chromatographic methods. Harada and Fox (1965b) used polyphosphoric acid as

Table 6.2 Syntheses of polypeptides by means of phosphoric acid derivatives

Monomers	Method	Reaction products	References
Gly	Trimetaphosphate + *N*-Heterocycles	Di- and triglycine	Katchalski and Paecht (1954)
D,L-Ala-Gly-Gly-OH	P_2O_5 in dimethyl phosphite	Oligopeptides up to DP 24	Schramm and Wissmann (1958)
Ala, Gly, Gln, Lys, Val + excess Asp	Polyphosphoric. a. 100 °C	Polypeptides with Mn up to c. 15,000	Harada and Fox (1965a)
Gly, Gly_2	Trimetaphosphate	oligoGly	Chung et al. (1971)
Ala, Pro	Mixed anhydrides with MP	Oligopeptides with DPs up to 5	Paecht-Horowitz and Katchalski (1967)
Ala, Asp	Mixed anhydrides with AMP + montmorillonite	PolyAla up to DP 54	Lewinson et al. (1967), Paecht-Horowitz et al. (1970), Paecht-Horowitz and katchalski (1973), Paecht-Horowitz (1974, 1976, 1978)
Ala, Asp, Gly, Ser	Activ. With ATP + montmorillonite	Various oligopeptides	Paecht-Horowitz (1984), Paecht-Horowitz and Eirich (1988)
Gly, His, Arg	Polyphosphoric acid + heat (100 °C)	PolyGly up to DP 24 His and Arg trimers	Sponer et al. (2024)

condensing agent for mixtures of various prAAs rich in Asp at temperatures around 100 °C. After dialysis, proteinoids having Mn′s up to 15,000 were obtained, but those authors did not check, if Aps was racemized. Therefore, these experiments were not much useful and informative.

In the following years, Paecht-Horowitz and coworkers explored another approach. At first, mixed anhydrides of prAAs and adenylic acid having protonated amino groups were prepared and after liberation of the amino group polycondensation of these activate amino acids was performed (Scheme 6.1). The activated derivatives of four different prAAs were studied, but their polycondensation in solution ended up with formation of low oligomers. A significant step forward was achieved by addition of montmorillonite. With this additive significantly higher molecular weights were achieved, for example, poly(L-alanine)s with DP's up to 54 (not isolated but identified in chromatograms), (Lewinson et al. 1967; Paecht-Horowitz 1974, 1976, 1978, 1984; Paecht-Horowitz and Katchalski 1973; Paecht-Horowitz et al. 1970; Paecht-Horowitz and Eirich 1988). Those studies and results represent an interesting and excellent performance from the scientific point of view, but for their relevance as arguments in favor of CE the following aspects need consideration. First, formation of significant concentration of such monomers under prebiotic conditions is absolutely unlikely. Successful polycondensation in low concentration is impossible, and an identification of compartments favoring synthesis and high concentrations of such activated prAAs in the presence of suitable minerals does not exist. Furthermore, copolypeptides having DPs >10 were not synthesized. Therefore, the Paecht-Horowitz experiments, although representing an excellent piece of scientific research, don't help to explain, how proteins/enzymes composed of more than 100 prAAs were formed in the course of CE. Almost 40 years later (in 2024), Sponer et al. reported on the synthesis of polyglycines up to a DP of 24 upon thermal polycondensation of Gly phosphoric acid salts However in the case of His and Arg, only trimers were obtained. Since longer chiral

Scheme 6.1 Synthesis of oligopeptides by polycondensation of mixed anhydrides of amino acids and adenylic acids

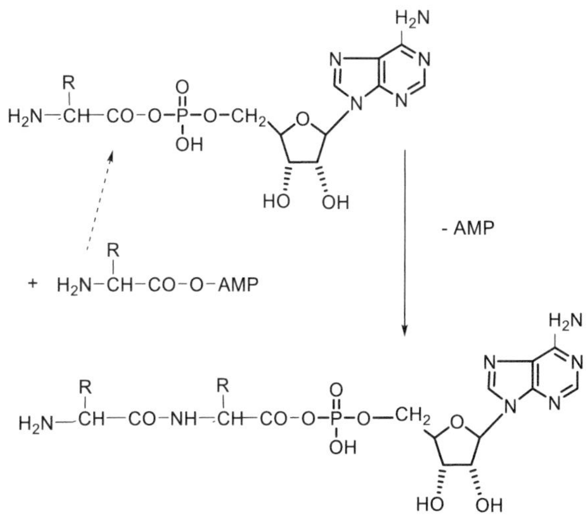

polypeptides were not obtained, the question of racemization was not discussed. Further critical arguments concerning all model syntheses of polypeptides will be discussed at the end of this chapter.

6.3 The "Carbonyl-Diimidazole Approach"

A third approach to a prebiotic synthesis of (co)polypeptides is based on carbonyl-diimidazole (CDI) as condensing agent. This approach was manly elaborated by Orgel and coworkers (Table 6.3). Their first report (Ehler and Orgel 1976) was focused on studies of the polymerization mechanism and yielded the reaction pathway outlined in Scheme 6.2. CDI does not promote a polycondensation process, it transforms the prAAs into the corresponding N-carboxyanhydrides, which in turn, start a s. c. CCP as discussed in Sect. 3.2. A similar mechanistic study was also reported in the same year by Ehler alone. Furthermore, Ehler et al. reported that CDI promoted polymerizations of Gly entail formation of small quantities of 2,5-diketopiperazines, the quantity of which considerably increases, when adenylic acid is added to the reaction mixture. In a subsequent publication, Ehler et al. (1977) confirmed the intermediate formation of prAA NCAs by the observation that Ser and Thr did not polymerize, but formed the transformation products of their NCAs formulated in Scheme 3.1.

Table 6.3 Polymerizations of prAAs by means of carbonyl diimidazole (CDI)

Monomers	Method	Reaction products	References
Ala, Gly	Imidazole buffer at 0 °C	Oligopeptides up to DP 5	Ehler (1976), Ehler and Orgel (1976)
Gly	CDI + AMP	Diglycine, 2.5 diketopiperazine	Brack et al. (1976)
Ser, Thr, His	Imidazole buffer at 30 °C	Oligomers of His, no olig. of Ser and Thr	Ehler et al. (1977)
Asp, Glu, O-Phos-Ser	Catalysis via Mg ions or cationic micelles	OligoAsp up to DP 13 OligoGlu up to DP 20 Oligo Ser up to DP 9	Hill Jr and Orgel (1996), Böhler et al. (1996)
Asp, Glu	Repeated add. of CDI + minerals	PolyGlu up to DP 50	Ferris et al. (1996)
Asp. Glu, O-Phos-Ser	Repeated add. of AAs to excess CDI + minerals	PolyAsp up to DP 27 PolyGlu up to 45 PolySer up to DP 13	Hill Jr et al. (1998)
Asp, Glu, α-Amino adipic a.	CDI or EDAC[a]	Oligomers up to DFP 10 with CDI only	Liu and Orgel (1998a)
Arg	CDI + various minerals	OligoAsp up to DP 10 with illite or FeS_2	Liu and Orgel (1998b)
Glu	Conc. NaCl solution	OligoGly up to DP 14	Wang et al. (2005)
Glu	Conc. NaCl or KCl solution	OligoGly up to DP 9 with KCl	Dubina et al. (2013)

[a]1-Ethyl-3-(3-dimethylaminopropyl)carbodiimide

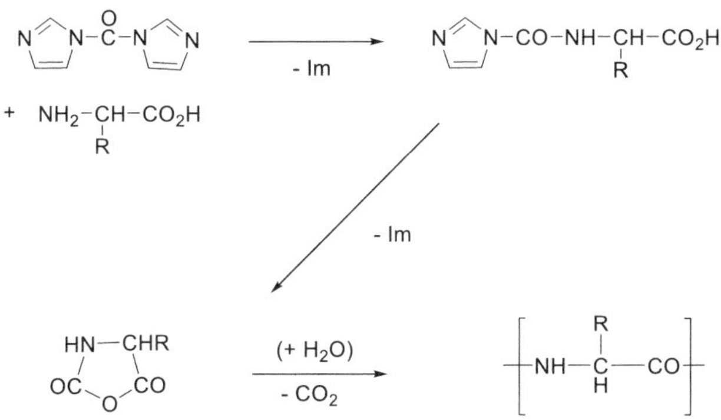

Scheme 6.2 Polymerization of amino acids by means of carbonyldiimidazole

About 20 years later, polypeptide syntheses by means of CDI were again studied by the Orgel group with focus on water-soluble prAAs such as Aps, Glu, and O-phosphorylated Ser (Hill Jr and Orgel 1996; Hill Jr et al. 1998; Böhler et al. 1996). Whereas the highest DPs of the earlier experiments of Ehler were limited to values around 5 only, the new experiments yielded higher oligomers, and when magnesium ions were added, DPs in the range of 10–20 were achieved. With addition of minerals such as illite or hydroxyapatite, poly(Glu) with DPs up to 50 were identified in chromatograms. Unfortunately, this success was limited to polymerizations of Glu and with other prAAs, the DPs remained below 20 under identical conditions. Liu and Orgel (1998a, 1998b) concentrated their experiments again on α-amino dicarboxylic acids and used minerals as catalytic additives hoping that adsorption of the intermediately formed NCAs favors formation of higher molecular weights Yet, DPs above 20 were not achieved. When CDI was replaced by water-soluble carbodiimide substantial amounts of 2,5-diketoüpiperazines were obtained. A positive result was the synthesis of poly(Arg) with DPs up to 10 in the presence of FeS_2. The experiments of Wang et al. (2005) and Dubina et al. (2013) were focused on the question, if and to what extent concentrated solution of NaCl or KCl can support the CDI promoted polymerization of Glu. Dubina et al. found that KCl was more favorable for the polymerization than NaCl, but DPs >10 were not achieved.

These CDI-based polymerization experiments represent certainly an interesting academic working field but it should be kept in mind that for several reasons, those experiments are not relevant for the hypothesis of CE. First, CDI is sensitive to hydrolysis and reacts within a few minutes with ammonia and primary amines. Second, there are no model experiments explaining how significant amounts of imidazole were formed under prebiotic conditions. Third, even in the presence of imidazole, formation of CDI cannot be explained in the neighborhood of concentrated prAA solutions. Fourth, several prAAs cannot be polymerized with CDI, for example, Ser, Thr, Cys, His, Lys. Fifth, the model experiments of the Orgel group were performed in the absence of chain terminators such as ammonia, carboxylic acids, or aldehydes.

6.4 Miscellaneous Methods

In this section and in Table 6.4, those model syntheses of oligopeptides are summarized that were not included in the afore-mentioned polymerization methods. The various methods presented in Table 6.4 may be grouped into two categories: first, methods based on reactive dehydrating agents (activation of amino or carboxylic groups); second, methods performed in concentrated salt solution without addition of activating agents. This latter method is called "*salt-induced polycondensation*" (SIP).

The earliest experiments based on the first approach (Ponamperuma and Peterson 1965; Steinmann et al. 1965; Steinmann 1967; Hawker and Oro 1981) have in common that cyanamide or biscyanodiamide were used as condensing agents. Yet, despite broad variation of the reaction conditions, DPs >4 were not obtained. With COS as condensing agent similarly poor results were obtained (Lehman and Orgel 2004). Using a water soluble carbodiimide, Liu and Orgel (1998a) achieved a DP of 6 and after addition of hydroxyapatite a DP of 10. These results confirm those experiments of Paecht-Horrowitz (Table 6.2), those experiments of Ferris et al. (1996)) or Hill Jr et al. (1998) and also those experiments of Liu and Orgel (1998a) with CDI in that addition of certain minerals has a considerably positive effect on the chain growth of oligopeptides. Weber and Orgel (1978, 1979a, 1979b, 1980, 1981) conducted numerous experiments with activated esters of Gly. They took into consideration that in living organisms, the biosynthesis of proteins involves condensation of activated ribose esters of the prAAs. However, in the model experiments of the Orgel group, DPs beyond 4 were not achieved, presumably due to the association of the oligoglycines.

The second approach, the SIP method, was explored by B. M. Rode and coworkers (Rode and Schwendinger 1990; Rode 1999). To ease the progress of the polycondensations, copper complexes of prAAs were used as monomers, but again, a DP of 4 was the limit which was never surpassed.

In the following years, five research group studies the polycondensation of Gly and Ala under various reaction conditions. Cleaves et al. (2009) and Otake et al. (2011) explored the influence of high temperatures and high pressure in the absence of catalysts in analogy to submarine hot spots. At temperatures up to 250 °C, the maximum DP was limited to 4, and at higher temperatures only a progress to DP 5 was achieved, but in addition to condensation steps, deamination and decarboxylation were also observed and Cleaves et al. concluded:

> Given the expected low concentration of amino acids, the higher residence times and the range of temperature in submarine hydrothermal systems (SHS) it is unlikely that SHS environments were robust sources of even simple peptides

Sauer et al. (2021) observed that the coordination of amino acids with Cu^{2+} ions favors condensation, but only low yields of dipeptides were obtained. Higher DPs were achieved by the same research group (Sydow et al. 2022) when metallic iron (delivered by meteors) and acetic acid were used as condensing agent in liquid SO_2.

Table 6.4 Syntheses of oligopeptides by means of various condensation methods

Monomers	Method	Reaction products	References
Gly, Leu	Cyanamide + UV	Tri- and tetrapeptides	Ponamperuma and Peterson (1965)
Ala, His, Leu, Phe	Cyanamide + ATP (80 °C)	Oligopeptides up to DP 4	Hawker and Oro (1981)
Ala Glu, Leu	Dicyandiamide	Di-and tripeptides	Steinmann et al. (1965), Steinmann (1967)
Phe	COS + H_2O	Di + tripeptides	
Aps, ß-Amino acids	EDAC[a]	Oligoamodes up to DP 6	Liu and Orgel (1998a)
Asp, ß-Amino glutaric a.	EDAC[a] + Illite or Hydroxy apatite	Oligoamides up to DP 10	Liu and Orgel (1998b)
Gly	Gly methyl ester + Cu^{2+} in solution	OligoGly up to DP 9	Brack et al. (1975)
Gly	Activ. as thioester of *N*-acetyl Cys	OligoGly up to DP4	Weber and Orgel (1979a)
Gly	Active. As 2′,3-ester of AMP or GMP	Diglycine + 2.5-diketopiperazine	Weber and orgel (1978, 1981)
Gly + 12 prAAs	Con. NaCl solution + Cu^{2+}	Dipeptides + traces of tri-and tetrapeptides	Weber and Orgel (1979b)
Ala, Asp, Gly, Glu, Val	Cu^{2+} catalyzed polycondensation in conc. NaCl solution (SIP method)	Dipeptides and traces of tri- and tetrapeptides	Schwendinger and Rode (1989, 1991, 1992), Schwendinger et al. (1995), Rode and Schwendinger (1990), Rode (1999), Saetia et al. (1993), Eder and Rode (1994), Suwannachot and Rode (1998)
Gly, Leu, Phe,Val	Heat + active. Al_2O_3	Mainly dipeptides	Bujdak and Rode (2001, 2002, 2003)
Gly	High pressure, high temp.(up to 250 °C)	OligoGly up to DP 4	Cleaves et al. (2009)
Gly, Ala	High press. (5.5 GPa) + high temp. (400 °C)	OlioGly + Oligo Ala up to DP 5	Otake et al. (2011)
Gly, Ala	Cu-catalysis in liquid SO_2	Dipeptides in yields <5%	Sauer et al. (2021)
Gly, Ala	Fe + acetic a. in liquid SO_2	OligoGly up to DP 10	Sydow et al. (2022)
Gly	Boric a. as catalyst	PolyGly up to DP 30	Sumie et al. (2023)

[a] 1-Ethyl-3-(3-dimethyl aminopropyl)carbodiimide

Even better results were obtained by the Furukawa group (Sumie et al. 2023) when boric acid was used as catalyst. The average DP reached a value around 15 and traces of polyglycines up to 30 were detected by MALDI TOF mass spectrometry. Yet, those authors did not explain, where the large amounts of boric acid required for their syntheses have existed on the early Earth. On the modern Earth, boric acid is an extremely rare mineral (called Sassolin). The metal salts are slightly more abundant, but they are not active as catalysts. Furthermore, it should be considered that even a successful synthesis of polyglycine is not relevant for the CE, if the synthetic method does not allow for the generation of copolypeptides containing trifunctional prAAs.

Finally, a publication of D. H. Lee et al. (1996) with the fascinating title "As self-replicating peptide" merits discussion. That publication begins with the statement:

> The production of α-amino acids and their condensation to polypeptides under plausible prebiotic conditions have long been known

As shown above and discussed below (Sect. 6.4), this statement is far from the truth. Furthermore, the experiments of Lee et al. do not deal with polycondensations of prAAs, they describe the coupling of two oligopeptide fragments one of which is composed of 15 AA units and the other one of 17 AA units. The 15 AA peptide carries a free amino and group and a CO–NH_2 group at the other end. The 17-AA peptide has an amino protecting group and a CO chain end activated as thio-benzyl ester. Hence, coupling yielding a peptide composed of 32 AA units is the only possible reaction. Nonetheless, side reactions were observed. Lee et al. have then found that addition of a small amount of the preformed 32-AA peptide catalyzed the coupling process and reduced the extent of side reactions. Since the new 32-AA peptide has the same catalytic activity as the added catalyst an autocatalytic acceleration takes place. This is, of course, an important scientific result, but for several reasons, it does not help to explain how and why CE yielding proteins under prebiotic conditions have occurred. First of all, such a template effect does not exist for the synthesis of proteins from monomeric amino acid derivatives. Second, there is no experimental evidence how all the 20 prAAs were formed under prebiotic conditions. Third, an efficient and selective synthesis of prAA thio-esters under prebiotic conditions has not been reported yet. The success of the Lee coupling method requires perfect homochirality of all prAAs. Fifth, even when all these problems have been solved, a successful polycondensation according to the Lee template effect is in contradiction to the RNA world concept. The final statement in the "Lee paper "is also worth to be cited:

> Given that amino acids and polypeptide species could have been produced under prebiotic conditions, the possibility of self-replicating peptide sequences should be considered in the early evolution of living systems.

Sentences, such as this one, have the tendency to convince readers with little knowledge in peptide chemistry, little knowledge in polycondensation theory, and little knowledge in RNA world chemistry that the problem of protein syntheses under prebiotic conditions has almost been solved. However, a new promising dimension in the field of polypeptide syntheses was opened in 2022 by a publication of F. Müller et al. (2022). Those authors reported that certain RNAs containing non-canonical nucleobases were capable to promote activation and polycondensation of prAAs in aqueous solution and played the role precursors in the CE toward a genetic code-controlled protein synthesis in ribosomes. Those authors presented the following final conclusion:

> All these data together support the idea that non-canonical vestige nucleosides in RNA have the potential to create peptide self-decorating RNAs and hence an RNA-peptide world. The formed RNA-peptide chimeras are comparatively stable, and so it is conceivable that some of these structures learned, at some point, to activate amino acids by adenylation and to transfer them onto the ribose OH-group to capture the reactivities in structure that were large and hydrophobic enough to exclude water. This would then have been the transition from the non-canonical nucleoside-based RNA-peptide world to the ribosome-centered translational process that is a hallmark of all life on Earth today.

6.5 Discussion of the Model Syntheses

At first it should be mentioned that a few researchers hailed the oligomerization of HCN as precursor of a prebiotic synthesis of peptides and proteins. The reason for this keen view was the finding that after hydrolysis of the oligomeric HCN, a few prAAs, mainly Gly, were chromatographically detected (Matthews and Moser 1966; Moser et al. 1968). However, the reaction conditions were so far from a plausible prebiotic scenario that a detailed discussion of that work is obsolete.

When all the afore-mentioned attempts to prepare polypeptides and proteins are considered and evaluated, fortunately a clear picture may be drawn. Even after 60 years of intensive research, all attempts to find a pathway yielding copolypeptides and proteins under plausible prebiotic conditions in the absence of RNAs have totally failed. This negative vote is based on five criteria:

1. Most experiments did not yield copolypeptides having DPs >10.
2. In a few cases where DPs >10 were achieved, the condensation method was far from plausible prebiotic conditions (e.g., use of CDI or large amounts of boric acid).
3. Syntheses of trace amounts of polyGly or polyAla with DPs above 10 are not acceptable as evidence for a prebiotic formation od proteins consisting of up to 20 different ürAAs with DPs around or above 100.
4. In the case of thermal polycondensations (Fox–Harada) unrealistic composition of the reaction mixture, thermal degradation of certain prAAs (Thr, Cys) and strong racemization prevent acceptance of the method.
5. Whenever metal complexes of prAAs or oligopeptides are involved, one has to consider that the rate of racemization is considerable higher (s. Sect. 10.3)

In summary, these condensation experiments confirm what has already been concluded from the model syntheses of prAAs. All results are diametrically opposed to a reductionistic-deterministic concept of CE. This conclusion has a positive aspect for the advocates of the RNA world. At first, evolution of RNAs incl. Species containing non-canonical nucleosides was necessary followed by combination and cooperation of certain RNAs in such a way that a machinery capable of synthesizing proteins having enzymatic properties became feasible. However, such a hypothesis makes the requirements for an evolution more stringent. It does not suffice to explain how a variety of short and long RNA chains was formed under prebiotic conditions. One has also to explain, how and why a genetic code for the 20 prAAs was developed. Furthermore, one has to explain why and how the various types of RNA have learned to construct a machine capable of synthesizing proteins with useful enzymatic properties. Such an evolution represents a complex process which proceeds with a loss of entropy in contrast to the chemistry going on in space (s. Chap. 9). Who or what has taught the prebiotic biomolecules to pursuit this way? Such a concept is not compatible with the hypothesis that live emerged on Earth on the basis of an undirected evolution. However, the assumption of a directed CE is in contradiction to the current mainstream interpretation of the biological evolution, and it requires the detection of analogous directed CEs on other planets to allow for a justification by universal law of nature. An origin of a directed CR as a unique creational act on Earth is outside the reach of scientific methods and scientific theories.

Finally, it should be emphasized the lack of all 20 prAAs and the lack of proteins in the primordial soup is deadly blow for the hypothesis that the origin of life is based on a unique lucky event, which brought all components necessary for the creation of the first living cell together. This hypothesis of the "fortunate coincidence" (Chap. 1, hypothesis II) is clearly in contradiction to numerous model experiments and in contradiction to the chemistry which proceeds in space (Chap. 9).

References

Balbiano L (1900) Ber Dtsch Chem Ges 33:2323
Balbiano L (1901) Ber Dtsch Chem Ges 34:1501
Böhler C, Hill Jr A, Orgel L (1996) Orig Life Evol Biosph 26:539
Brack A, Ehler K, Orgel L (1976) J Mol Evol 8:307
Brack A, Louembe D, Spach G (1975) Orig Life 6:407
Bujdak J, Rode B (2001) Amino Acids 21:281
Bujdak J, Rode B (2002) Inorg Biochem 90:1
Bujdak J, Rode B (2003) J Therm Anal Chem 73:397
Chung N, Lohrmann R, Orgel L (1971) Tetrahedron 21:1205
Cleaves H, Aubrey A, Bads J (2009) Orig Life Evol Biosph 39:109–126
Curtius T, Benrath A (1904) Ber Dtsch Chem Ges 37:1270
Dubina M, Vyazmin S, Boitsov V, Niklaev E, Popov I, Kononikin A, Elisev I, Natochin I (2013) Orig Life Evol Biosph 43:109
Eder A, Rode B (1994) J Chem Soc Daölton Trans 1994:1125
Ehler K (1976) J Org Chem 41:3041
Ehler K, Girard E, Orgel L (1977) Biochim Biophys Acta 491:253
Ehler K, Orgel L (1976) J Mol Evol 8:307
Ferris J, Hill Jr A, Liu R, Orgel L (1996) Nature 381:59

Fox S, Dose K (1977) Molecular evolution and the origin of life. 2nd edn. Marcel Dekker, New York
Fox S, Harada K (1960) J Am Chem Soc 82:3745
Fox S, Harada K, Rohlfing D (1962) In: Stahmann A (ed) Polyamino acids, polypeptides and proteins. Univ. of Wisconsin Press, Madison, p 5
Fox S, Middlebrook M (1954) Fed Proc 13:211
Fox S, Vegodsky A, Harada K, Hogland P (1957) Ann N Y Acad Sci 69:328
Harada K (1959) Bull Soc Chem Soc J 32:626
Harada K, Fox S (1958) J Am Chem Soc 80:2694
Harada K, Fox S (1965a) Arch Biochem Biophys 109:49
Harada K, Fox S (1965b) In: Fox S (ed) The origins of prebiological systems and of their molecular matrices. Academic Press, New York
Hawker J, Oro J (1981) J Mol Evol 17:285
Heinrich M, Rohlfing D, Bugna E (1969) Arch Biochem Biophys 130:441
Hill Jr A, Böhler C, Orgel L (1998) Orig Life Evol Biosph 28:235
Hill Jr A, Orgel L (1996) Orig Life Evol Biosph 26:539
Katchalski A (1951) Adv Protein Chem 6:123
Katchalski A, Paecht M (1954) J Am Chem Soc 76:600
Krampitz G (1962) In: Stahmann M (ed) Polyamino acids, polypeptides and proteins. Univ. of Wisconsin Press, Madison, p 5
Lee D, Granja J, Martinez J, Severin K, Ghadin M (1996) Nature 382:525
Lehman L, Orgel L, Ghadin M (2004) Science 306:283
Lewinson R, Paecht-Horowitz M, Katchalski A (1967) Biocheim Biophys Acta 140:24
Liu R, Orgel L (1998a) Orig Life Evol Biosph 28:47
Liu R, Orgel L (1998b) Orig Life Evol Biosph 28:245
Matthews C, Moser R (1966) Proc Natl Acad Sci USA 56:1087
Meggy A (1954) Chem Soc 1954:1444
Moser R, Chagett A, Matthews C (1968) Tetrahedron Lett 13:1005
Müller F, Escobar L, Xu F, Wegrzyn E, Nainyte M, Amatov T, Chan C-Y, Pichler A, Carell A (2022) Nature 605:279
Otake T, Taniguchi T, Furukawa Y, Kawamura F (2011) Astrobiology 11:799
Paecht-Horowitz M (1974) Orig Life Evol Biosph 5:173
Paecht-Horowitz M (1976) Orig Life Evol Biosph 7:369
Paecht-Horowitz M (1978) J Mol Evol 11:101
Paecht-Horowitz M (1984) Orig Life Evol Biosph 14:307
Paecht-Horowitz M, Berger J, Katchalski A (1970) Nature 20:636
Paecht-Horowitz M, Eirich F (1988) Orig Life Evol Biosph 18:359
Paecht-Horowitz M, Katchalski A (1967) Biochim Biophys Acta 140:14
Paecht-Horowitz M, Katchalski A (1973) J Mol Evol 2:91
Ponamperuma C, Peterson E (1965) Science 147:1572
Rode B (1999) Peptides 29:773
Rode B, Schwendinger M (1990) Orig Life Evol Biosph 20:401
Saetia S, Riedel K, Eder A, Rode B (1993) Orig Life Evol Biosph 23:167
Sauer F, Haas M, Sydow C, Siegle A, Lauer C, Trapp O (2021) Nat Commun 12:7182
Schramm G, Wissmann H (1958) Chem Rev 91:1073
Schwendinger M, Rode B (1989) Anal Sci 5:441
Schwendinger M, Rode B (1991) Inorg Chim Acta 186:247
Schwendinger M, Rode B (1992) Orig Life Evol Biosph 22:349
Schwendinger M, Tauler R, Saetia S, Kromer R, Rode B (1995) Inorg Chim Acta 228:207
Sponer J, Culon R, Olyepka M, Sponer J, Siegle A, Trapp O, Siepokura K, Zdrahal Z, Sedo O (2024) Commun Chem 7:185
Steinmann G (1967) Arch Biochem Biophys 121:533
Steinmann G, Lemmon R, Calvin M (1965) Science 147:1574
Sumie Y, Sato K, Kagegawa T, Furukawa Y (2023) Nat Comun Chem 6:89

Suwannachot Y, Rode B (1998) Orig Life Evol Biosph 28:79
Sydow C, Sauer F, Segle A, Trapp O (2022) ChemSysChem. https://doi.org/10.1002/syst.202200034
Wang K, Yao N, Li C (2005) Orig Life Evol Biosph 35:313
Weber A, Orgel L (1978) J Mol Evol 11:189
Weber A, Orgel L (1979a) J Mol Evol 13:193
Weber A, Orgel L (1979b) J Mol Evol 13:385
Weber A, Orgel L (1980) J Mol Evol 16:1
Weber A, Orgel L (1981) J Mol Evol 17:190
Wills C, Bada J (2000) The spark of life. Perseus Publications, Cambridge

Model Syntheses of Oligo- and Polynucleotides

7

It requires more courage to change his point of view than to maintain it.

Norisat Persehkian

7.1 Various Condensation Methods

The first attempts to synthesize oligo or poly(ribonucleotide)s (Table 7.1) were published shortly before 1960 (Khorana et al. 1956; Tener et al. 1958; Michelson 1959). A. M. Michelson polymerized the 2′,3′-cyclophosphates of all four nucleosides by means of phenyl phosphorochloridate (Scheme 7.1) and obtained mixtures of 2′,5′and 3′,5′connected nucleoside polyphosphates (Scheme 7.2a). This author also described a stepwise synthesis up to a DP of 6. Khorana and coworkers (Khorana et al. 1956, 1957; Tener et al. 1958; Coutsogeorgupopulos and Khorana 1964) used *N,N′*-dicyclohexylcarbodiimide as condensing agent and avoided the problem of reactions at the 2′-OH group by using thymidylic acid (a 2′-deoxynucleoside) as monomer (Scheme 7.3a). In a later publication (1964), they blocked the 2′-OH group by acetylation and used the 3′-phosphates as monomers for polycondensation (Scheme 7.3b). Oligo(nucleotide)s up to a DP of 6 were isolated, and oligo(nucleotide) s up to DP 11 were identified by chromatography.

G. Schramm and coworkers studied numerous polycondensation experiments using nucleoside −5′monophosphates as monomers and phosphoric anhydrides as condensing agent (Schramm et al. 1961, 1962; Schramm 1965). These phosphoric anhydrides were used, because theses condensing agents might have existed on the early Earth in contrast to the coupling agents mentioned before. However, Hayes and Hansbury reported only a few month later that the Schramm method does not yield longer polynucleotide chains, and the question if and to what extent the 2′-OG group was involved remained unanswered. Polyphosphoric acid was also used by

H. R. Kricheldorf, *Life by Chemical Evolution?*
https://doi.org/10.1007/978-3-662-70752-4_7

Table 7.1 Syntheses of oligo- and polynucleotides via polycondensation by means of various condensation agents or via ring-opening polymerization of cyclophosphates

Starting materials	Method (cond. Agent)	Products	References
AMP, GMP, CMP, UMP, TMP	Diphenyl phosphorochloridate or tetraphenyl pyrophosphate	Oligonucleotides	Michelson (1959)
Adenosine diethyl pyrophosphate	Heat	Oligoadenylic acid	Cramer (1961)
AMP, CMP, GMPTMP, UMP	Ethyl metaphosphate	Polynucleotides	Schramm et al. (1961), Schramm (1965)
TMP	Ethyl metaphosphate	Complex polynucleotides	Hayes and Hansbury (1964)
CMP	Polyphosphoric acid	Oligo- and poly(cytidylic acid)	Schwartz et al. (1965)
TMP	Dicyclohexylcarbodiimide + Pyridine	Cyclic and linear oligo(nucleotides)	Khorana et al. (1956, 1957), Gilham and Khorana (1958), Tener et al. (1958)
2′-Acetyl-uridyl-5′-phosphate	Dicyclohexylcarbodiimide + Pyridine	Oligo(uridylic acid), up to DP 10	Coutsogeorgupopulos and Khorana (1964)
TMP	Cyanamide + Montmorillonite	Cyclic + linear oligonucleotides	Ibanez et al. (1971)

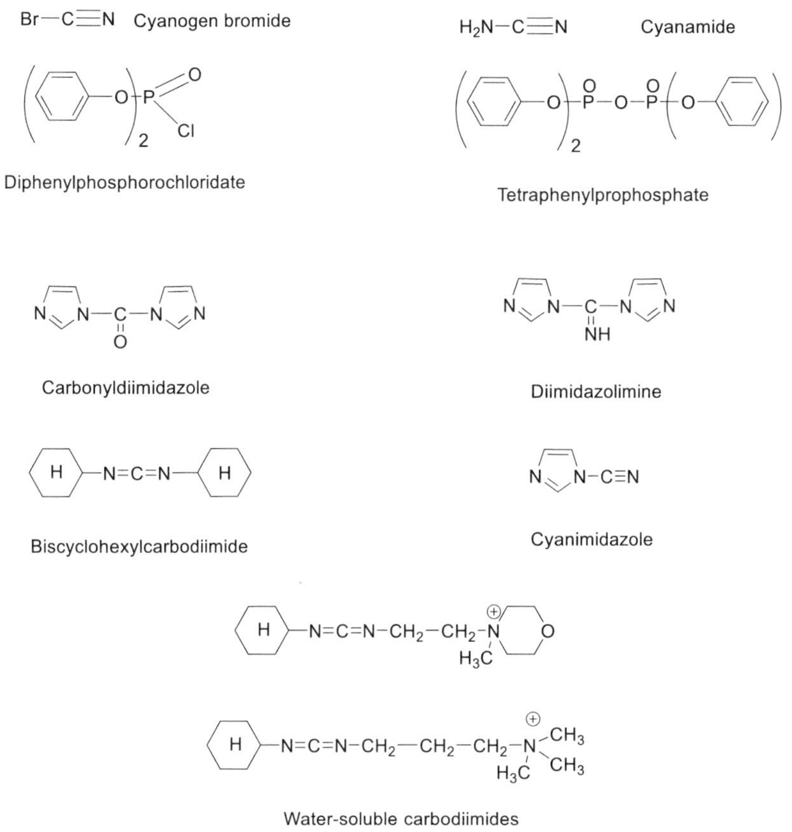

Scheme 7.1 Dehydrating (condensing) agents used for the synthesis of oligonucleotides

Fox and coworkers (Schwartz et al. 1965) with similarly negative results. Oligomers were only obtained from cytidylic acid, and the yields extremely low.

After 1967, the research group of L. E. Orgel began with systematic polycondensation studies (Schneider-Bernloehr et al. 1968; Sulston et al. 1968a, b). AMD was mainly used as monomer, and water-soluble carbodiimides served as condensing agents (s. Scheme 7.1). A first stimulating result was the observation that addition of poly(uridylic acid), poly(U), accelerated the polycondensation process and enhanced the yields. When mixtures of two or three different nucleotides were used in combination of poly(U), a high selectivity of the incorporation of AMD was found, but dimers were the main reaction products.

The condensation of 2′-deoxyAMP in the presence of polyU yielded the pyrophosphate as main product, and the reaction of AMP with adenosine yielded the 5′,5′-phosphoric acid diester.

It is, of course, absolutely unlikely that water-soluble carbodiimides were present on the early Earth, but cyanamide, the isomer of unsubstituted carbodiimide, was certainly present, in as much as this molecule also belongs to the "space

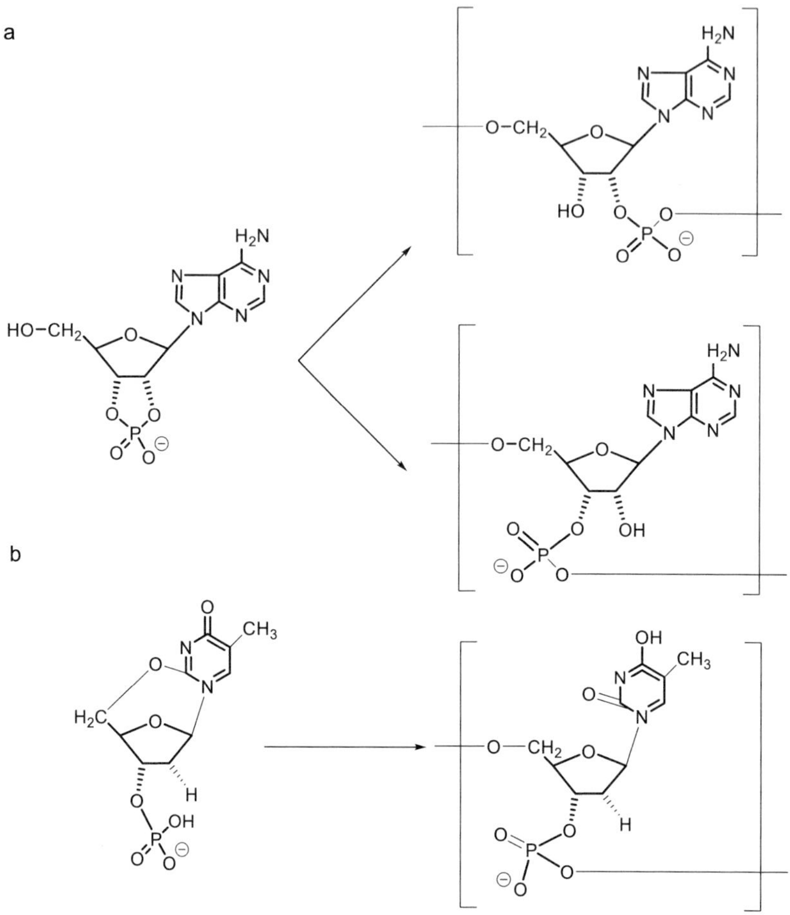

Scheme 7.2 Examples of nucleic acid syntheses by ring-opening polymerization

chemicals." For this reason, Ibanez et al. (1971) studied the polycondensation of thymidylic acid by means of cyanamide. Since it is derived from 2′-deoxyribose, the problem of isomeric chain growth steps does not exist. These experiments are remarkable, because higher oligomers were only formed in the presence of montmorillonite and because formation of cyclic oligonucleotides was detected.

Again starting out from thymidine, Nagyvary and Nagpal (1972) explored the ring-opening polymerization of O^2–5′cyclothymidine-3′-phosphate (eq. (b) in Scheme 7.2). These polycondensations were performed simply by heating without addition of a catalyst or template. Surprisingly this simple procedure yielded oligomers up to a DP of 12, and seemingly 3′-5′-linkages were exclusively formed.

To achieve higher molar masses, several research groups used preformed oligonucleotides (tetra-, penta-, and-hexanucleotides) as starting materials (quasi monomers) (Table 7.2). This synthetic strategy has, of course, the advantage that a few

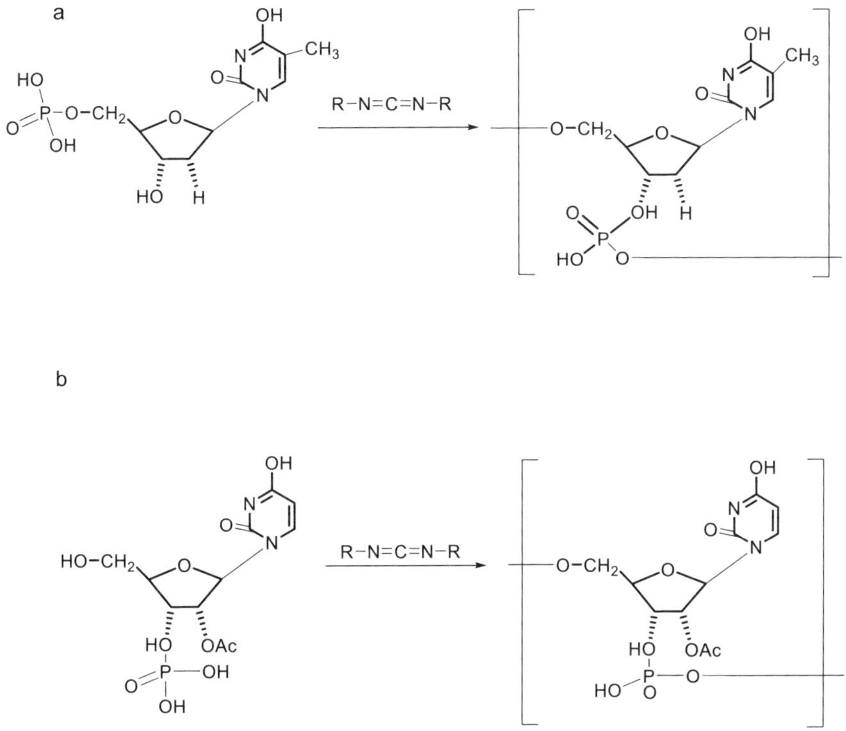

Scheme 7.3 Regioselective syntheses of nucleic acids using carbodiimides

coupling steps suffice to reach DPs above 20. First experiments in this direction were already reported in 1966 by Naylor and Gilham (1966). Those authors dimerized penta- or hexa-thymidylic acid along a poly(A) template by means of a water-soluble carbodiimide. Unfortunately, only yields around 3–5% were obtained. A water-soluble carbodiimide was also used by Uesugi and Tso (1974) for polycondensations of the 2′-methylinosine. These unusual monomers were synthesized by cautious hydrolysis of 2′-methylated poly(inosinic acid) and contained one chain end in the form of a phosphorylated 3′-OH group. In the presence of a poly© template, up to five coupling steps were realized, so that polynucleotides having DPs around 30 were obtained.

Only two or three coupling steps were achieved by Usher and McHale (1976) when hexa (A) having 2′0.3′-cyclophosphate end group was coupled along a poly(U) matrix. Again 1,2-diamino ethane served as catalyst, and the dodecamer was isolated in a yield of 24% but for the octadecamer, the yield was only around 5%. Considerably higher yields and higher DPs were achieved by Yanagawa using a hexa(A) having a 5′-phospahte end group as "monomer" in combination with a poly(U) template. Bromocyanide in combined with imidazole served as condensing agent, and the catalytic effect of various bivalent metal ions. By means of Ni^{2+}, the highest selectivity for 3′,5′-linkages was found and yields up to 60% were achieved.

Table 7.2 Coupling of preformed oligonucleotides (cited in timely order)

Starting materials	Method	Products	References
Penta(T) or Hexa(T)	Poly(A) + water-sol. carbodiimides	Deca(T) or Dodeca(T)	Naylor and Gilham (1966)
Oöigo(2′-O-methylinosine-3′-phosphate)	Templates + water-sol. carbodiimides	Oligonucleotides	Uesugi and Tso (1974)
Hexa(A) with cyclic phosphate end group	Poly(U) + ethylene diamine	Dodeca(A) + Octadeca(A)	Usher and McHale (1976)
Hexa(A)	Poly(U) + bromocyanide + metal ions	Poly(A) up to DP 48	Kanya and Yanagaw (1986)
3′-Aminodinucleotide	Templates + water-sol. carbodiimides	Tetranucleotides with 3′,5′-phosphoamide linkage	Zielinski and Orgel (1987a, b)
Oligonucleotides hydrol. from DNA or RNA	Template + water-sol. carbodiimide	Double helix segments	Dolinnaya et al. (1988)
Oligonucleotides hydrol. from DNA or RNA	Bromocyanide or water-sol. carbodiimides	Double helix segments	Dolinnaya et al. (1991)
Oligonucleotides	Water-sol. carbodiimides	Higher oligonucleotides	Harada and Orgel (1994)
Nucleoside triphosphates	Template + metal ions	Oligonucleotides	Rohatgi et al. (1996a)
Various oligonucleotides	Templates, Na-metaphosphate + Mg^{2+}	Dimers or trimers of the oligonucleotides	Gao and Orgel (2000)
D(CGTA)	*N*-Cyanoimidazole + ethidium bromide	Cyclic oligomers + polynucleotides up to DP 100	Horowitz et al. (2010)
2′-acetylated nucleotides	Template + cyanoimidazole	Higher oligonucleotides with high percentage of 3′,5-linkages	Bowler et al. (2013)

The highest DP was 48 AMP units indicating eight successful coupling steps. Cyanoimidazole and diimidazole imine were identified as reaction intermediates (Scheme 7.1). A detailed discussion of the usefulness of various condensing agent for the coupling of oligonucleotides was published by Ferris in 1989 (Ferris et al. 1989). Cyanoimidazole was again the condensing agent in the work of Horowitz et al. (2010), who studied the polycondensation of tetranucleotides along the complementary matrix. DPs up to 100 nucleotide units were achieved. This conspicuous success was achieved by addition of multicyclic heterocycles, such as proflavine or Coralyn, which intercalate into the matrix and in the growing oligonucleotides, so that a stiffening effect occurs, which prevents cyclization of the oligonucleotides. In the absence of these intercalating agents, cyclic tetra- and cyclic octa-nucleotides were the main reaction products. Other authors, who reported on syntheses of oligonucleotides, did not mention the formation of cyclic oligomers either because they did not expect their formation and/or because they did not have the analytical methods for their detection. However, the low yields reported in most publications suggest that cyclization was an important side reaction in most studies. Anyway, the results reported by Horowiz et al. indicate that the modern theory of polycondensation (s. Sect. 3.1) is, of course, also valid for syntheses of biopolymers.

Several research groups investigated the coupling of two oligonucleotides in the presence of DNA segments. The oligonucleotides were prepared by enzymatic degradation of DNA or RNA chains. However, high molecular weights were not the goal of this work, but a comparison of the strength of association between the oligomeric "monomers" and natural matrices, or the preparation of double helices with well-defined defects (Harada and Orgel 1994; Rohatgi et al. 1996b; Dolinnaya et al. 1988, 1991; Bowler et al. 2013). Bowler et al. used a selective acetylation of the 2′-OH group to achieve a perfect 3′/5′coupling. Finally, the work of Zielinski and Orgel (1987a, b) merits detailed discussion. Those authors intended to develop a self-replicating poly(nucleotide) synthesis. The success of that work was not achieved with true nucleotides as monomers but with analogous 3′-amino-3′-deoxynucleotides, the condensation of which yielded phosphor-amide groups (Scheme 7.4). By means of the tetramer GNHpCNHpGNHpCNHp, those authors realized the coupling of PGNHpCN3 with GNHpCNH2. The new tetramer was itself active as matrix and catalyzed the synthesis of the parent matrix, and in this way, an autocatalytic polycondensation process was realized. Yet, those authors did not succeed to establish such an autocatalytic process for normal (oligonucleotide) s, and thus, the synthetic procedure explored by Zielinsky and Orgel cannot be accepted as a realistic reaction pathway for the chemical evolution.

In this connection, a review article of Hulshof and Ponamperuma (1976) should be cited. In that article, those authors compared the usefulness of various condensing agents in the presence of water under the aspect of their availability on the early Earth. The following agents were considered: carbodiimides, cyanamide, dicyanodiamide, HCN-tetramer, and several cyclic or linear polyphosphates, and the following conclusion was drawn: "From both, a chemical as well as biological point of view the polyphosphates appear to be the most plausible general prebiotic condensing agents."

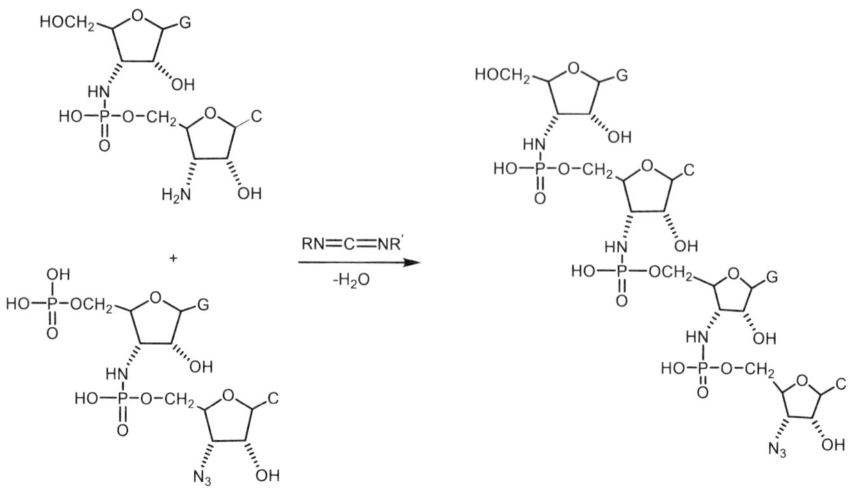

Scheme 7.4 Regioselective synthesis of poly(3′-amino-3′-deoxyribonucleotides) using water soluble carbodiimides

Finally, the synthesis of nicotinamide adenine dinucleotide (NAD), a prosthetic group of redox enzymes, should be mentioned. The synthetic approach, published by the Trapp group (Bechtle et al. 2024), is particularly remarkable, because it started out from primitive precursor molecules such as ammonia, cyanoacetaldehyde, prop-2-ynal and sugar forming starting materials.

7.2 Syntheses of Oligonucleotides Via Nucleotide Imidazolides

The most widely studied method for the polycondensation of nucleotides with respect to prebiotic RNA syntheses is based on activation of the phosphate group in the form of imidazolides (or methylimidazolides) followed by suitable polycondensation procedure (Scheme 7.1). The hitherto published methods may be subdivided into the following three groups:

1. Polycondensation in the absence of a matrix (template)
2. Polycondensation supported by an inorganic template
3. Polycondensation in the presence of an oligo- or polynucleotide template

The methods listed in Table 7.2 follow the afore-given order and will be discussed in this order. A first series of detailed experiments concerned polycondensations of AMP imidazolides or (rarely) UMP imidazolides in water catalyzed by various heavy metal ions (Table 7.3), (Sawai 1976; Sawai and Orgel 1975; Sawai et al. 1981, 1989; Sleeper and Orgel 1979). Whereas the addition of most metal ions had little effect, addition of Zn^{2+} enhanced the yield by a factor of 10 (Sawai and Orgel 1975). A further increase of yield and DP (up to a DP of 10) was achieved by addition of Pb^{2+}ions, and it was found that the solubility of the lead salt had an

Table 7.3 Model syntheses of polynucleotides via the imidazolide method

Starting materials	Method, template	Reaction products	References
ImpA, imp(A)$_2$, impU	Polycond. in $H_2O + Zn^{2+}$, Pb^{2+}, or UO_2^{2+}	OligoRNA with 2′,5′-linkages; DP up to 10 with metal ions	Sawai and Orgel (1975), Sawai (1976), Sawai et al. (1981, 1989), Sleeper and Orgel (1979)
ImpA, meimpC, meimpG	Polycond. in the presence of montmorillonite	OligoRNA, DP up to 4	Ferris and Ertem 1992, 1993), Ferris et al. (1996), Ertem and Ferris (1996)
ImpA, impG, impT, impU, impC	Polycond. with poly(U) or poly(C)	Homodinucleotides, DP up to 4	Wejmann et al. (1968)
ImUpG, impCpA	Poly(A-C) and pol(G-U) as templates	Alternating poly-nucleotides, DP <10, mainly 3′,5′-linkages	Ninio and Orgel (1978)
ImpG	Poly(C) as template. Influence of Zn^{2+} was studied	OligoG, DFP <10, yields up to 75%, Zn^{2+} favors 3′,5′linkages	Bridson and Orgel (1980) van Roode and Orgel (1980), Fakrai et al. (1981)
ImpG	Poly(C) + Pb^{2+}	Poly(G), DP <17, yield up to 85%, mainly 2′,5′-linkages	Lohrmann and Orgel 1980)
ImpA + various isomers	No template	Selective polycond. of D-ribosides	Lohrmann and Orgel (1977)
ImpApA, impC-C	Poly(U) or poly(C) used as templates	Altern. oligonucleotides, 2′,5′- and 3′,5′-linkages	Lohrmann and Orgel (1979a, 1979b), Lohrmann et al. (1981)
ImpA	Poly(U) + hydroxyapatite	Oligo(A)	Gibbs et al. (1980)
ImpA, impU	Precipit. Pb salts of nucleotides as templates	Oligonucleotides, DP <6	Orgel and Lohrmann (1974), Lohrmann (1982)
MeimpA, meimpG, meimpC	Poly(C), poly(G) or CpCpGpCpC as templates	G/C Oligomers	Inoue and Orgel (1982), Rembold and Orgel (1994)
MeimpA, meimpG, meimpC	Random poly(AC), poly(C/G), poly(G/C) as templates	Oligo/A/G) and oligo(G/C) with random sequences	Joyce et al. (1984b), Joyce and Orgel (1986)
MeimpG + meimpU	Random poly(C/A) as template	Random oligomers, DP <13	Joyce and Orgel (1988)
MeimpG + MeimC	Hepta(G-C) as template	Altern. oligonucleotides	Nig and Orgel (1989)
ImpG + MeimpG + MeimpA	Template + deca(C) as primer	Chain extension by 6–9 nucleotides	Rodriguez and orgel (1991)

(continued)

Table 7.3 (continued)

Starting materials	Method, template	Reaction products	References
L- and D-MeimpG	Poly(C)	L-monomers hinder polycond. Of D-monomers	Joyce et al. (1984b)
D-MeimpG	Plly(C) containing 1 or 2 L-ribose units	L-Ribose in template chains do not hinder polycond.	Kozlov et al. (1998)

important influence on its catalytic efficiency (Sawai 1976; Sawai et al. 1981; Sleeper and orgel 1979). Yet, the highest yield (95%), the highest DP (12) were achieved with addition of UO_2^{2+} ions. (Sawai et al. 1989). All these experiments were conducted without addition of a template and had in common that predominantly or even exclusively the undesired 2′,5′-connection was formed.

The experiments performed by J. P. Ferris and G. Ertem had in common that all polycondensations were conducted in the presence of montmorillonite (Ferris and Ertem 1992, 1993; Ferris et al. 1996; Ertem and Ferris 1996). In the first part of their work, ImpA served as monomer (Scheme 7.4) and DPs up to 10 were identified by chromatographic methods. In contrast to the experiments of the Orgel group, addition of montmorillonite favored the formation of the desired 3′,5′-phosphate diester. In the following part, the influence of various metal ions was examined, and Mg^{2+} or Al^{3+} were found to be ineffective. Montmorillonite doped with UO_2^{2+} ions favored the formation of 2′,3′-phosphate linkages in analogy to the results reported by Sawai et al. (1989). With a CMP activated by 2-methyl imidazole, Ferris et al. (1996) synthesized oligo(C)s up to a DP of 15. With polycondensation of AMP 2-methyl imidazolide, a record DP of 55 was achieved, when in regular time intervals addition of the monomer was several times repeated. Those authors did not discuss what happened to the large excess of monomer, but hydrolysis and cyclization reaction were the most likely side reactions that consumed a large part of the monomer.

First experiments dealing with polycondensations of nucleotide imidazolides in the presence of a polynucleotide template were reported by the Orgel group (Wejmann et al. 1968). AMP imidazolide was polycondensed in the presence of poly(U) at 0 °C. After long reaction times, the dimer and the trimer were detected. No information about the extent of 2′,5′-phosphate groups was disclosed. In the years 1978–1981, the imidazolides of all five nucleotides relevant for RNA and DNA were synthesized and polycondensed in the presence of the complementary templates. Ninio and Orgel (1978) found that the pyrimidine monomers ImpU and ImpC alone did not polymerize along poly(A) or ply(G). Either dimers or trimers were needed to stabilize the base pairing with the matrix. Dimers of pyrimidine and purine bases (e.g., ImpUpG, or ImpCpA) reacted well with the complementary alternating templates.

Bridson and Orgel (1980), van Roode and Orgel (1980), and Fakrai et al. (1981) explored the influence of Zn^{2+} ions on polycondensations of ImpC in the presence of poly(C) and found a considerable increase of both yields and DP. Particularly remarkable was the large predominance of 3′,5′-linkages. With addition of Pb2+, Lohrmann and Orgel (1980) obtained even slightly higher yields and DPs (up to

17), but as already found by Sawai et al. (Sawai and Orgel 1975; Sawai et al. 1981, 1989) in the absence of a template, the formation of 2′,5′-linkages was now prevailing.

In several publications, the same authors reported on various aspects of ImpA and ImpApA polycondensations in the presence of a poly(U) matrix (Lohrmann and Orgel 1977, 1979a, b). Perhaps the most interesting results were obtained, when polycondensations of ImpA were compared with analogous imidazolide monomers derived from variants of the A structure in the presence of poly(U). Such variants were, for example, α-D-adenosine, 2′- or 3′-deoxyasenosine, and ara-adenosine. Regardless which variant was present, the fastest reaction was that of ImpA with its own species, what means that the polycondensation of ß-D-ribose nucleotides is highly selective, when a suitable matrix is present. This is certainly a remarkable result clearly supporting the RNA world hypothesis. Further studies of Lohrmann and Orgel revealed that precipitates of lead salts containing nucleotides acted as weak templates for the oligomerization of nucleotide imidazolides (Orgel and Lohrmann 1974; Lohrmann 1982). Therefore, those authors uttered the hypothesis that such precipitates might have acted as precursor of oligonucleotide templates in the beginning of the molecular evolution.

In a publication of Inoue and Orgel (1982), it was for the first time demonstrated that nucleotides activated by 2-methylimidazole showed a higher efficiency in coupling reactions than the normal imidazolides. Using MeimpG as monomer (Scheme 7.5) and poly(C) as matrix, they were able to realize syntheses of poly(G)s having DPs up to 50 in combination with yields up to 80%. Therefore, the Orgel group

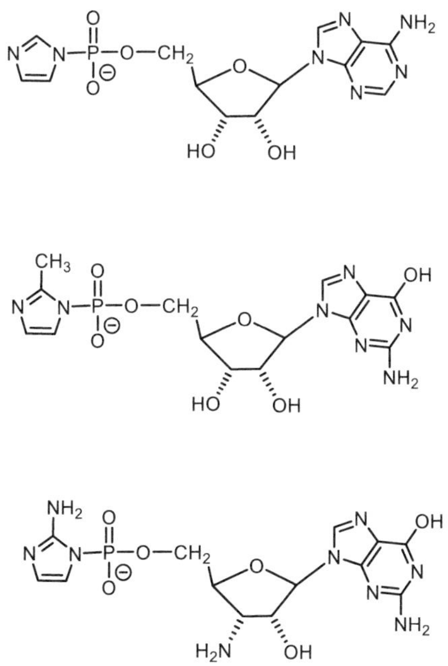

Imp A

Meimp G

Imp G NH2

Scheme 7.5 Structures and acronyms of selected nucleotides activated as imidazolides

continued their studies preferentially with methylimidazole activated nucleotides. However, this kind activation requires a pH around 8 as optimum reaction condition. Among the numerous (poly)condensation studies published in the years 1984–1998, the publications of Joyce et al. (1984a) and Kozlov et al. (1998) merit to be highlighted. Those authors discovered that addition of L-MeimpG significantly hampered polycondensations of D-MeimpG along a poly(C) template consisting of D-ribose units. In contrast, one or two L-units implanted in the poly(C) chain did not hinder their function as template, but only reduced its efficiency. This important finding suggests the conclusion that the emergence of the RNA world was only possible in the absence of L-ribose based nucleotides. The prebiotic synthesis of D-saccharides under the influence of optically active catalysts based on L-a-methyl amino acids delivered by meteorites provides a speculative explanation for the existence of such a scenario. However, this scenario also requires that the reaction conditions prevented rapid racemization of the D-ribosides. Unfortunately, no information is available about the stability of D-ribose under the reaction conditions yielding D-saccharides Further studies of Kozlow et al. should be mentioned: Kozlow et al. (1999a, b; 2000).

Several research groups studied template-promoted oligonucleotide syntheses using chemically modified templates or modified Meimp nucleotides. For example, Vischer et al. (1989) synthesized a poly(C) matrix containing pyrophosphate groups between the D-ribose units. Hill et al. (1991) used normal ply(C) or poly(A) chains as templates but used Meimp(3-isoG) or Meimp(3-isoxanthine) as monomers. Toiichi et al. (1987) prepared oligo(phosphor-ester-amides) of nucleosides by means of 2′-amino- or 3′-amino 2′-deoxy or 3′-deoxynucleotides activated in the form of imidazolides.

After 2003, the group of J. W. Szostak launched a research program based on the polycondensation of nucleotide imidazolides. The numerous experiments summarized in Table 7.4 may be subdivided into three groups which also represent a timely order. Group 1 is characterized by "amino- monomers" such as 2′-amino- or 3′-amino, 2′-deoxy or 3′-deoxy saccharides in analogy to the experiments by Toiichi et al. Since the amino group is considerably more nucleophilic than a hydroxy group, these oligomerization reactions involved a high degree of regioselectivity. The resulting poly(phosphor-ester-amide)s were studied with regard of their base-pairing properties, for example, by comparing the stability of their double helices with double-stranded RNA. The group 2 experiments were performed with artificial variants of nucleotide imidazolides, but neither primers nor templates were used to support the polycondensation process. The purpose of those experiments was to get more information about the reaction mechanisms and to optimize the reaction conditions. The theoretical background of the group 1 and group 2 experiments was the experience that even the template supported oligomerization of normal nucleotide imidazolides is relatively slow process involving side reaction. The group 3 experiments starting out from the experience of the Orgel group that activation of nucleotides with 2-methyl imidazole is more effective than that with imidazole itself. The Szostak group began to explore the useful ness of 2-aminoimidazole as activating agent and found a further progress in terms of higher reaction rates and yields.

Table 7.4 Syntheses of RNAs via the imidazolide method by Szostak and coworkers

Starting materials	Method	Products	References
2′-Amino-2′-deoxyribonucleotides, 3′-Amino-3′-deoxyribonucleotides. 2′-amino-2′,3′-deoxyribonucleotidees	Meimp-activated polycondensations with and without addition of templates	Oligomeric phosphor-amide-esters, Kinetic studies	Zhang et al. (2012, 2013a, b), Schrum et al. (2009), Blain et al. (2014), Izgu et al. (2016)
Various ribonucleotides	Template supported, primer-initiated polycond. of Meimp nucleotides	Oligomeric RNAs	Zhang et al. (2012), Engelhart et al. (2013), Adamale et al. (2015), Pal et al. (2016), Zhang et al. (2017), Giurgiu et al. (2017), Tan et al. (2017), Jia et al. (2016, 2017)
Various ribonucleotides	Template supported polycond. of nucleotides activated with 2-aminoimidazole	Oligomeric RNAs	Walton and Sczostack (2017), Li et al. (2017), Fahrenbach et al. (2017)

Particularly remarkable is the finding that a prebiotic plausible reaction pathway for the formation of nucleotide(2-aminoimidazolides) was discovered. Whereas a prebiotic formation of imidazolides or 2-methylimidazolides is extremely unlikely (Fahrenbach et al. 2017).

More recently, enzyme-free, but template supported coupling of mono-or dinucleotide imidazolides has found increasing interest (Hey et al. 2003; Duzdevich et al. 2021). The coupling of dinucleotide imidazolides proved to be significantly more efficient than coupling of mononucleotides.

In this connection, it should be mentioned that even more than 20 years ago, speculations were forwarded that the activation of nucleotides resulted from phosphorylation of nucleobases, above all *N*-methyladenine, because the formation of imidazolides was and is highly unlikely (Prabahr and Ferris 1997; Huang and Ferris 2003):

> Since adenine and other purines are required for the synthesis of RNA-like monomers, it is likely that adenine or its derivatives would be present in sufficient amounts on the early Earth to also serve as phosphate activating groups. 1-Methyladenine is one of the reaction products of methylamine with adenine….

R. Shapiro has published a negative response to this speculation in 2006 (Shapiro 2006). Nonetheless, the question remains unanswered, why N-heterocycles such as benzimidazole (Scheme 5.3) or unusual purine and pyrimidine bases did not serve as activators of nucleotides. Finally, a review article dealing with non-enzymatic RNA syntheses in the presence of RNA templates should be mentioned (Kozlow and Orgel 2000).

7.3 Ring-Opening Polymerization (ROP) of Cyclic Nucleoside Phosphates

An interesting synthetic approach which is not based on condensation steps is the ROP of cyclic phosphates (Table 7.5). In the early 1970s, the group of L. E. Orgel explored ring-opening polymerizations of various nuclseoside-2′,3′-cyclophosphates (Eq. (a) in Scheme 7.2). In a first publication (Renz et al. 1971), the influence of various basic catalysts on the formation of a AMP dimer (ApAp) was studied in the presence of a poly(U) template. In the following papers (Verlander et al. 1973; Verlander and Orgel 1974), polycondensations of adenosine-2′-3′-cyclodiphosphate without template in bulk were described. 1,2-diaminoethane served as catalyst, and in the optimum case, a DP of 13 was achieved, but the yield was below 1%. Traces of higher oligonucleotides were detected by chromatographic methods. Despite this remarkable success, about 30 years later, the DeMauro group (Constanzo et al. 2009, 2012; Pino et al. 2013) explored the ROP of 3′,5′-cGMP. It was found that this energy rich monomer can be polymerized without addition of templates and without addition of enzymes simply by heating in conc. Aqueous solution or in polar organic solvents such as formamide or dimethyl formamide. Temperatures in the range of 40–90 °C were found to be useful, and a highly selective formation of 3′,5-phosphate linkages was observed. Tertiary amines had a catalytic effect, and under optimized conditions poly(guanylic acid) with DPs up to 25 were obtained. Ligation of poly(G) with the complementary poly(C) was also observed. Work on ROP of 3′,5′-cGMP was resumed by the group of D. Braun

Table 7.5 Ring-opening polymerizations of cyclic D-ribose phosphates

Starting materials	Method	Products	References
O2, 5′-Cyclo-thymidylic-3′-phosphate	Ring-opening polymerization (ROP)	Ologo(thymidylic acid) up to DP 12	Nagyvary and Nagpal (1972)
Adenosine-2′,3′-cyclophosphate (cAMP)	ROP with and without addition of Poly(U)	Oligo(adenylic acid) up to DP 5	Renz et al. (1971), Verlander et al. 1973; Verlander and Orgel 1974
3′,5′-cGMP	Non-enzym. ROP in various solvents	Poly(guanylic a.) up to DP 25	Constanzo et al. (2009, 2012), Pino et al. (2013)
3′,5′-cGMP		Poly(guanylic a.)	Morasch et al. (2014) Sponer et al. (2015)
2′,3′-cUMP, cCMP, cAMP, cGMP	ROP at pH 7–12 in bulk or air/water Interf. No cat.	Oligo(guanylic a.) up to DP 10	Dass et al. (2023)
2′,3′-cCMP, cGMP	ROPs in heated artif. rock cracks	Oligo nucleotides	Dischert et al. (2023)
2′,3′-cA, caG, cU or cC monophosphates	ROP in slightly basic aqueous solution catalyzed by prAAs	Oligonucleotides	Rout et al. (2024)

(Morasch et al. 2014; Sponer et al. 2015). Long strands of polynucleotides were also obtained by ROPs in the dry state. In combination with terminal ligation, RNA chains with DPs up to 120 nucleotide units were achieved.

More recently, the Braun group focused their activities on ROPs of 2′,3′-cyclo-phosphates, because the work of Powner and Sutherland (e.g., Powner et al. 2009, s. Chap. 5) opened a new promising access to a direct prebiotic synthesis of 2′,3′-cyclo-phosphates without needing free D-ribose as starting material. Although the 2′,3′-cyclophosphate group is thermodynamically more stable and less reactive than the 3′,5′-rings, spontaneous ROPs were found by drying at air–water interfaces (Dass et al. 2023) at temperatures in the range of 40–80 °C, whereupon DPs up to 10 were achieved. Heated rock cracks were found to concentrate nucleotides and to favor ROPs of 2′, 3′-cyclophosphates (Dischert et al. (2023). A new interesting aspect came into play when Rout et al. (2024) discovered that prAAs catalyze the ROP of 2′-3′-cyclophosphates upon heating in the dry state at pHs in the range of 8–10, whereupon DPs up to 10 were reached. Those authors wrote:

> Here we show that amino acids substantially improve the yield of RNA polymerization from prebiotic plausible ribonucleoside-2′,3′-cyclic phosphates. The amino acid catalysis of RNA polymerization was base selective, increased the compositionally oligomer sequence diversity and favored the formation of natural 3′,5′-linkages. It was most efficient atpH values close to the pH_{aH} of the amines ad correlated with side chain hydrophobicities, suggesting a general acid-base mechanism akin to ribozyme-catalyzed RNA hydrolysis. The copolymerization ofG/C/A/U with valine resulted in a substantially increased oligomer yield and compositionally diversity. The catalytic effect of prAAs on RNA oligomerization suggests the reactivity of amino acids and RNA was united earlier than previously thought, and their association likely predates the generic code.

7.4 Discussion of the Model Syntheses of Oligonucleotides

The oligo- and polynucleotide syntheses presented and discussed above may be subdivided into three groups:

1. Polycondensations by means of phosphoric acid anhydrides (incl. Polyphosphoric acid) as condensing agents
2. Polycondensations by means of organic condensing agents such as carbodiimides
3. Polycondensations of nucleotide imidazolides
4. Polymerizations of cyclic 2′,3′-nucleoside phosphate

Phosphoric anhydrides regardless of low or high molecular weight are sensitive to hydrolysis, and their existence in a world largely covered by water over long periods of time is extremely unlikely. On the other hand, one can exclude that poly-phosphates were repeatedly generated in special environments in combination with volcanic activities. Yet even, when this possibility is conceded, one has to consider that the results obtained with model syntheses were pretty poor in three directions: first, the DPs were low; second, a significant regioselectivity favoring 3′,5-phosphat linkages did not exist; third, pyrophosphate groups were incorporated into the grow-ing polynucleotide chains. Furthermore, it was assumed but not clearly proven that ß-D-ribosides were the largely prevailing repeat units. Concerning the experiments with cyanobromide, carbodiimides, and nucleotide imidazolides, it is necessary to emphasize that those model syntheses are far from the prebiotic reaction conditions regardless if a primordial soup or an evolution primarily proceeding in compartments is considered. Any evidence is lacking that those reagents were present on the early Earth in relevant amounts.

The only reaction pathway which is acceptable as model of the prebiotic evolution of RNAs is the ROP of cyclic phosphates. The direct synthesis (avoiding free D-ribose as reaction partner) of nucleotide 2′,3′-cyclophosphates and their ring-opening polymerization as reported by Powner and Sutherland (e.g., Powner et al. 2009, s. Chap. 5) provides a sound experimental background for this hypothesis above all for 2′,3′-cyclophospahtes. The first studies of ROPs with cyclic nucleophosphates were disappointing (Verlander et al. 1973; Verlander and Orgel 1974), because DPs above 13 were only found in yields around 0.15%, but the recent studies of ROPs of nucleotide cyclophosphates by the D. Braun group (2022, 2023, 2024, Table 7.5) are more promising. However, it must be taken into account that the ROP approach based on cyclophosphates of nucleotides requires for successful syntheses of RNAs that it evolved in compartments or in mineral layers (for reasons of monomer concentration) and it requires the absence of nucleotides derived from L-ribose. Concerning the concentration of nucleotides and short RNAs on certain mineral surfaces and in rock cracks, recent studies of Dischert et al. (2023) and Helmbrecht et al. (2023) should be mentioned.

Another new working field of the Braun group concerns the emergence of structured sequences by replication via template ligation Toyabe and Braun 2019; Endeleva et al. 2019; Kudella et al. 2021; Rosenberg et al. 2021). Finally a review article should be mentioned (Sponer et al. 2016) which is focused on "Emergence of the first Catalytic Oligonucleotide in a Formamide-Based Origin Scenario."

References

Adamale T, Engelhad A, Sczostak J (2015) J Am Chem Soc 137:483
Bechtle M, Kurrle N, Trapp O (2024) Chem Eur J 30:e202402055
Blain J, Ricardo A, Sczostak J (2014) J Am Chem Soc 136:2033
Bowler F, Chan C, Duffy C, Garlaan B, Islam S, Powner F, Sutherland J (2013) Nat Chem 3:383
Bridson P, Orgel L (1980) J Mol Biol 144:567–577
Constanzo G, Pino S, Cicirello F, DiMauro E (2009) J Biol Chem 284:33206
Constanzo G, Saladino R, Botta G, Girorgi A, Scipioni A, Pino S, DiMauro E (2012) ChemBioChem 13:999
Coutsogeorgupopulos C, Khorana A (1964) J Am Chem Soc 86:1920
Cramer F (1961) Angew Chem 73:49
Dass A, Wunnawa S, Langlais J, von der Esch B, Kruche M, Ufer L, Chrisam N, Dubini R, Gartner F, Angerpointer S, Dischert C, Rovo P, Mast C, Sponer J, Ochesenfeld C, Frey E, Braun D (2023) ChemSysChem 5:e202200026. https://doi.org/10.1002/syst.2022000026
Dischert C, Ianeselli A, Tetiker D, Matreux T, Queener R, Mast C, Braun D (2023) Phys Chem Chem Phys 25:3375
Dolinnaya N, Sokolova N, Ashirkebova D, Shabarova Z (1991) Nucleic Acid Res 11:3067
Dolinnaya N, Sokolva N, Gryaznoya O, Shaqbarova Z (1988) Nucleic Acid Res 16:3721
Duzdevich D, Carr C, Ding D, Zhang S, Walton T, Szostack J (2021) Nucleic Acids Res 49:3681
Endeleva E, Salditt A, Stamp J, Schwintek P, Boekhoven J, Braun D (2019) RSC Chem Sci 10:5807
Engelhart A, Powner M, Sczostak J (2013) Nature 5:390

Ertem G, Ferris JP (1996) Nature 379:238
Fahrenbach A, Giurgiu C, Tam C, Li L, Hongo Y, Aono Y, Sczostak J (2017) J Am Chem Soc 139:7880
Fakrai H, van Rode J, Orgel L (1981) J Mol Evol 17:295–302
Ferris J, Ertem G (1992) Science 257:1387
Ferris J, Ertem G (1993) J Am Chem Soc 115:12270
Ferris J, Hill Jr A, Liu R, Orgel L (1996) Nature 381:59
Ferris J, Huand CH, Hagan WJ (1989) Nucleosides Nucleotides 8:407
Gao K, Orgel L (2000) Orig Life Evl Biosph 30:45
Gibbs D, Lohrmann R, Orgel LE (1980) J Mol Evol 15:347–357
Gilham PT, Khorana HG (1958) J Am Chem Soc 80:6212
Giurgiu C, O'Flaherty DK, Tann C, Sczostak JW (2017) J Am Chem Soc 139:16741
Harada K, Orgel L (1994) J Mol Biol 38:558
Hayes F, Hansbury E (1964) J Am Chem Soc 86:4172
Helmbrecht V, Weingart M, Klein F, Braun D (2023) Geology 21:758
Hey M, Hartel C, Göbel M (2003) Helv Chim Acta 86:844
Hill A, Kumar S, Patil V, Leonard N, Orgel L (1991) J Mol Evol 32:447
Horowitz E, Engelhart A, Chen M, Quales K, Smith MW, Lyn D, Hud N (2010) Proc Natl Acad Sci USA 107:5288
Huang W, Ferris J (2003) Chem Commun 2003(12):1458
Hulshof J, Ponamperuma C (1976) Orig Life 7:187
Ibanez JD, Kinnball A, Oro J (1971) Science 173:444–446
Inoue T, Orgel L (1982) J Mol Evol 162:201–217
Izgu E, Oh SS, Sczostack JW (2016) Chem Commun 52:3684
Jia T, Fahrenbach A, Kanat K, Abdamale K, Sczostack J (2016) Nat Chem 8:915
Jia T, Fahrenbach AC, Kanat K, Abdamale K, Sczostack J (2017) Nat Chem 9:1286
Joyce G, Viser G, van Bockelt C, van Boom J, Orgel L, van Vestren J (1984a) Nature 310:602–604
Joyce G, Inoue T, Ortgel L (1984b) J Mol Biol 176:279–300
Joyce G, Orgel L (1986) J Mol Biol 188:433–441
Joyce GF, Orgel LE (1988) J Mol Biol 202:677–681
Kanya E, Yanagaw H (1986) Biochemistry 25:7423–7430
Khorana HG, Tener GM; Moffat JG; Pol RM (1956) Chem. Ind, 1523
Khorana HG, Razzeli WE, Gilliam P, Tenerr GM, Pol RM (1957) J. Am. Chem. Soc. 79:1002
Kozlov J, Pitsch S, Orgel L (1998) Proc Natl Acad Sci USA 95:13448
Kozlow IA, DeBouvere B, Van Aerschot A, Herdewijn P, Orgel LE (1999a) J. Am. Chem. Soc. 121:5856–5859
Kozlow IA, Politis PK, Pitsch S, Herdewijn P, Orgel LE (1999b) J. Am. Chem. Soc. 121: 1108–1109
Kozlow IA, Zielinski M, Allart B, Kerremans L, Van Aerschot A, Busson R, Herdewijn P, Orgel LE (2000) Chemistry Eur. J. 6:151–155
Kudella P, Tlachenko A, Salditt AY, Maslov S, Braun D (2021) PNAS 118(8):e2018830118. https://doi.org/10.1073/pnas2018830118
Li L, Prywes N, Tam C, O'Flaherty DK, Lilyveld V, Izgu E, Pal A, Sczostack J (2017) J Am Chem Soc 139:1810
Lohrmann R (1982) J Mol Evol 18:185–195
Lohrmann R, Bridson P, Orgel L (1981) J Mol Evol 17:303–306
Lohrmann R, Orgel L (1977) J Mol Biol 119:193
Lohrmann R, Orgel L (1979a) J Mol Evol 12:237–257
Lohrmann R, Orgel L (1979b) J Mol Evol 14:243
Lohrmann R, Orgel L (1980) J Mol Biol 142:555–567
Michelson A (1959) J Chem Soc (Lond) 1371:3655
Morasch M, Mast C, Langer J, Schilcher P, Braun D (2014) Chembiochem 15:879
Nagyvary J, Nagpal KL (1972) Science 177:272
Naylor R, Gilham P (1966) Biochemistry 5:3722
Nig K, Orgel L (1989) J Mol Evol 29:101
Ninio J, Orgel L (1978) J Mol Evol 12:91
Orgel L, Lohrmann R (1974) Acc Chem Res 7:365

Pal A, Das R, Zhang W, Lang M, McLaughlin M, Larry W, Sczostack J (2016) Chem Commun 52:11905
Pino S, Constanzo G, Giorgi A, Sponer J, Sponer JE, DiMauro E (2013) Entropy 15:5362
Powner M, Gerland B, Sutherland J (2009) Nature 459:230
Prabahr K, Ferris J (1997) J Am Chem Soc 119:4330
Rembold H, Orgel L (1994) J Mol Evol 38:205
Renz M, Lohrmann R, Orgel LE (1971) Biochim Biophys Acta 240:463–471
Rodriguez L, Orgel L (1991) J Mol Evol 33:477
Rohatgi R, Bartel D, Sczostack J (1996a) J Am Chem Soc 118:3332
Rohatgi R, Bartel D, Sczostack J (1996b) J Am Chem Soc 118:3340
van Roode J, Orgel L (1980) J Mol Biol 144:579
Rosenberg J, Göppel T, Kudella P, Braun D (2021) Am Phys Soc Phys Rev 11:0310565
Rout S, Wunnaya S, Powner M, Sponer J, Mast C, Braun D (2024) Res Square. https://doi.org/10.21203/rs.3.rs-4343389/v1
Sawai H (1976) J Am Chem Soc 98:7037
Sawai H, Kuroda K, Hojo T (1989) Bull Chem Soc (Jpn) 63:2218
Sawai H, Orgel L (1975) J Am Chem Soc 97:3532
Sawai H, Shibata T, Ono M (1981) Tetrahedron 37:481
Schneider-Bernloehr H, Lohrmann R, Sulston J, Weimann BJ, Orgel LE (1968) J Mol Biol 37:131
Schramm (1965) in The Origin of Prebiological Systems (Fox SW ed.), Academic Press, N. Y
Schramm G (1961) In: Fox SW (ed) The origin of prebiological systems. Academic Press, New York, p 299
Schramm G, Grötsch H, Pollmann W (1961) Angew Chem 73:619
Schramm G, Grötsch H, Pollmann W (1962) Angew Chem Int Ed 1:1
Schrum J, Ricardo A, Krishnamurthy M, Blain J, Sczostack J (2009) J Am Chem Soc 131:14560
Schwartz AW, Bradley E, Fox SW (1965) In: Fox SW (ed) The origin of prebiological systems. Academic Press, New York, p 317
Shapiro R (2006) Q Rev Biol 81:105
Sleeper H, Orgel L (1979) J Mol Evol 12:357
Sponer J, Sponer J, Novakoya O, Brabec V, Sedo O, Zdrahal Z, Costanzo G, Pino S, Saladino R, DiMauro E (2016) Chem Eur J 22:3572
Sponer JE, Sponer J, Giorgi A, DiMauro E, Pino S, Constanzo G (2015) J Phys Chem B 119:2979
Sulston J, Lohrmann R, Orgel L, Miles TH (1968a) Science 60:400
Sulston J, Lohrmann R, Orgel LE, Miles TH (1968b) Proc Natl Acad Sci USA 59:409, 726
Tan CFP, Zhu L, Fahrenbach AC, Zhang W, Walton T, Sczostack JW (2017) J Am Chem Soc 140:783
Tener G, Khorana H, Markham R, Pol E (1958) J Am Chem Soc 80:6223
Toiichi M, Zielinski WS, Chen C, Orgel L (1987) J Mol Evol 25:97
Toyabe S, Braun D (2019) Am Phys Soc Phys Rev 9:011056
Uesugi S, Tso P (1974) Biochemistry 13:3142
Usher DA, McHale A (1976) Science 192:53
Verlander M, Lohrmann R, Orgel L (1973) J Mol Evol 2:303–316
Verlander M, Orgel L (1974) J Mol Evol 3:115–120
Vischer J, Baker C, van de Woerd R, Schartz A (1989) Science 244:4902
Walton W, Sczostack JW (2017) Biochemistry 56:39
Wejmann B, Orgel L, Schneider-Bernloehr H, Sulston J (1968) Science 244:4902
Zhang N, Zhang S, Blain J, Sczostack J (2013b) J Am Chem Soc 135:924
Zhang N, Zhang S, Sczostack J (2012) J Am Chem Soc 134:3691
Zhang S, Blain J, Zielinski W, Gyraznov S, Sczostack J (2013a) Proc Natl Acad Sci USA 110:17732
Zhang W, Tam C, Walton W, Fahrenbach A, Birrane G, Sczostack J (2017) Proc Natl Acad Sci USA 114:7659
Zielinski W, Orgel L (1987a) Nature 327:127
Zielinski W, Orgel L (1987b) Nature 327:346

Further Reading

Kozlov J, Orgel L (2000b) Mol Biol 34:781–789

Copolymerization, Self-Reproduction, and Genetic Code

8

It is by experiment that we prove, but by intuition that we discover.

Jules H. Poincare

8.1 Copolymer Sequences and the Formation of Proteins

Proteins like polynucleotides are copolymers from the viewpoint of systematical polymer chemistry. Copolymers differ from homopolymers in three structural parameters:

1. They consist of two or more different types of monomeric units (comonomers).
2. The molar ratio of these comonomers may vary.
3. The sequence of the comonomers may vary.

If one ignores that a few enzymes contain the selenium analogs of cysteine, one may say that all proteins are made up of 20 different prAAs (and all RNAs contain the four different nucleobases U, C, A, G). The various biological functions of proteins do not result from a variation of the prAAs (comonomers) but from variation of their molar ratio (parameter 2) and of their sequences (parameter 3). An analogous conclusion also holds for RNAs. Consider a protein consisting of 200 prAA units and free combination of all 20 different prAAs, an undirected evolution can yield so many different proteins (billions), that, as discussed in Chap. 2, formation of significant amounts of those proteins, necessary for the first cell, is absolutely unlikely. The presence and incorporation of large amounts of non-protein AASs are an additional unsolved problem. The same complex problems also exist for the formation of RNAs, although less severe due to the lower number of nucleobases and related heterocycles that can interfere with nucleobases.

What can be said from the view point of theoretical polymer chemistry and from the view point of model experiments about the formation of different sequences

H. R. Kricheldorf, *Life by Chemical Evolution?*
https://doi.org/10.1007/978-3-662-70752-4_8

from a combination of two, three, or four different prAAs? About 50 years ago, a theory of copolymerization of two or three olefins or vinyl monomers was elaborated, taking into account a radical (irreversible) polymerization of comonomers having different reactivities (Elias 1997; Billmeyer 1984). Consider a binary system where the amounts of incorporated monomer A and monomer B were determined for various reaction times and various conversions. These data enable the calculation of kinetic parameters for the reaction between A and A, B and B, A and B, and B and A: k_{AA}, k_{BB}, k_{AB}, k_{BA}. On the basis of these rate constants, the following copolymerization parameters were defined: $r_A = k_{AA}/k_{AB}$ and $r_B = k_{BB}/k_{BA}$. These copolymerization parameters allow for the following simple information. When r_A and r_B are both larger than 1, the copolymerization will yield blocky sequences, but when both parameters are below 1, the copolmerization has a trend toward alternating sequences.

I. The copolymerization must obey the definition of chain-growth polymerization, and the condensative chain polymerization (CCP) defined in Sect. 3.2 belongs to this category. Furthermore, the syntheses of biopolymers in living organisms also belong to this category. However, polycondensations are step-growth polymerizations where monomers, oligomers, and polymers react with each other at any time, and thus, polycondensations do not allow for the determination of copolymerization parameters.
II. The polymerization process must be irreversible. For syntheses of biopolymers under prebiotic conditions, irreversibility may perhaps be assumed for temperatures below 25 °C, but not for temperatures above 70 °C at pH values <6 or >8, as they exist in the neighborhood of black or white smokers. Under harsh conditions (partial), hydrolytic degradation of biopolymers combined with efficient syntheses involves a tendency toward thermodynamically controlled random sequences.

Interestingly, copolymerizations of prAA NCAs obey both definitions; they are CCPs and they are irreversible. Therefore, copolymerizations of prAA NCAs are useful model reactions allowing one to check the usefulness of the afore-mentioned copolymerization theory for syntheses of copolypeptides. Furthermore, several combinations of prAAs have the advantage that ^{13}C NMR and ^{15}N NMR spectroscopy allow for a differentiation and quantification of A-A, A-B, B-A and B-B connections (dyads). Several studies in this direction were reported by Kricheldorf et al. (Kricheldorf and Schilling 1978; Kricheldorf and Hull 1980; Kricheldorf et al. 1985, 1986; Hull and Kricheldorf 1980). It was, for example, found that Ala NCA is more reactive than Val-NCA and thus, the incorporation of Ala units was significantly faster than that of Val units, with the consequence that blocky sequences were formed. Since the bulky side chain of Val (and Ile) reduces the reactivity of the NCA ring for steric reasons, this result was not unexpected. However, a surprising result was the finding that the composition of the secondary structure was to large extent kinetically controlled. Blocks of Ala units favor the formation of α-helices, whereas blocks of Val units prefer the ß-sheet structure. It was found that the α-helix/ß-sheet ratio measured after reprecipitation, and thus, under thermodynamic control differed from that fixed in the virgin samples via the kinetically controlled polymerization process. Hence, these model experiments illustrate what also living organisms frequently do, namely the formation of kinetically controlled secondary structures as consequence of the biological protein synthesis.

Beyond these results, the model reactions of NCAs did not give any answer on the question, why the canonical 20 prAAs were selected for the biosynthesis of proteins and why certain sequences were found to favor special biological functions. In older treatises about the origin of proteins and enzymes (Haldane 1929; Oparin 1924, 1938, 1957, 1961), a detailed discussion of these questions does not exist, because those authors did not have sufficient knowledge about the prebiotic synthesis of prAAs and about the failure to find an acceptable, trustworthy synthesis of polypeptides under prebiotic conditions. Even in more recent years, these difficult questions were rarely addressed. Exceptions are the books of Wills and Bada (2000), DeDuve (2002), and Luisi (2006). Wills and Bada wrote the following comment (p. 120):

> It is also fairly easy to explain why life today is based on only on the twenty amino acids commonly found in proteins. These amino acids have structures that lend themselves to formation of peptides in regular chains, with side groups that exhibit a diversity of different functions. Most of the many other amino acids that were made in the Miller-Urey experiment, as well as those found in meteorites such as Murchison, tend to have chemically similar and rather uninteresting side groups. Worse, they do not join together to form regular peptide chains. Chains formed from these amino acids would have had branched, jagged and clumpy structures. There must have been a great deal of selection among these early protein-like molecules for those that had sufficiently regularity to enable them to perform useful chemical tasks.

It is pretty difficult to outperform the superficiality and misleading character of this comment. For example: what is the meaning of "regular chain"? A regular sequence in the mathematical sense? No enzyme chain obeys such a kind of regularity. A relatively high extent of regularity does certainly exist in scleroproteins, such as collagen, keratin, or elastin. Yet, these proteins mainly consist of only three prAAs, Gly, Ala, and Pro, whereas prAAs functional side chains are pretty rare. In both, model experiments and in meteorites, non-protein AAs such as α-amino butyric acid, α-amino iso-butyric acid, and ß-alanine are pretty abundant, but is nonsense to say that incorporation of such AAs in polypeptides will result in a branched or clumpy character.

Those authors also mention that a fractionation of polypeptides having different compositions may have occurred upon repeated crystallization and solvation from aqueous solution in lakes and sea basins. This assumption is, in principle, correct, but it does not explain how a selection toward proteins with specific biological functions has taken place. Furthermore, neither any acceptable hypothesis nor any model experiments exist explaining how large copolypeptides and proteins were formed from solutions of prAAs under prebiotic conditions. Of course, one may assume on the basis of the RNA world hypothesis that at first a kind of RNA machinery emerged which in turn produced the desired proteins. But if so, Willis's and Bada's concept of selection via fractionated crystallization is obsolete.

DeDuve (2002) postulated that RNAs suitable as prAA carriers appeared first and played the role of transfer RNAs (p. 68):

> There is every reason to believe that proteins, as defined in the preceding chapter – polypeptides made from twenty specified kinds of amino acids – are an invention of RNA. Needless to say that the word "invention" is not intended here in its usual sense. What is meant is that the first true proteins arose from chemical interactions between RNA molecules and amino acids.

In agreement with the fact that DeDuve was a fan of the reductionistic-deterministic view of the CE (s. Sect. 1.1), he did not believe that such an interaction was based on chance events (p. 61):

> Without going into details of chemical structure, let it simply be said that the spontaneous genesis in some "primeval soup" of a molecular arrangement like RNA defies chemical common sense. Indeed, it has so far defied the ingenuity of chemists.

As a committed adherent of the RNA world hypothesis and of the reductionistic concept DeDuve was convinced that the precursors of transfer RNA were somehow formed in the course of the CE, but he was not able to explain how (p. 61):

> There is every reason to believe that the mergence of RNA was a crucial step in the development of life which preceded and most probably determined the appearance of DNA and of proteins, but before RNA there must have been something else that prepared and caused the advent of this key substance.

When a broad variety of RNAs existed, they had no other goal in mind than the rapid synthesis of biologically useful proteins (p. 69):

> Any model for the development of the protein synthesis must necessarily start with direct interaction between RNAs and amino acids. One does not see how else how RNA could ever have become involved in assembling amino acids. Most likely certain RNA molecules became linked with certain amino acids. This remember is exactly what happens today between transfer RNA and amino acids [a wonderful "a posteriori argument" instead of experiments]……Did the supposed association take place randomly between any kind of RNAs and amino acids? Such a selectivity would notably, provide an answer to a question many consider one of the great mysteries posed by the origin of life: why are proteins made from the 20 amino acid species that serve universally for their synthesis? It is not a question of relative abundance. Some amino acids present in large quantities, both in meteorites and in products of abiotic syntheses, are not used for protein assembly. Others, though rare, are [.....]. This explanation provides a simple answer to the riddle of the proteinogenic amino acids, even though it may not entirely account for the chirality problem.

These statements of DeDuve are examples that demonstrate how a strong belief in the RNA world in combination with a strong belief in the reductionistic-deterministic view overcomes any deficit of evidence-based knowledge and any difficulty to explain chemical miracles. Fortunately, the comments of Luisi are more realistic and critical (Luisi 2006, p. 65):

> Looking at the few illustrative data presented here on the polycondensation of amino acids into polypeptides [corresponds roughly to a short version of Chapter 6] one can conclude that a variety of interesting reactions has been proposed and studied, some of potential interest in the prebiotic scenario. However, the question of how to make long and specific sequences in a way relevant on the origin of life is still open. For example, there is very little in the literaturefor the synthesis and characterization of copolypeptide chains containing, say 30 residues, obtained under alleged prebiotic conditions (up to the spring of 2006). Generally, then, the question "why this and not that? – namely why a particular polypeptide sequence was formed, and not a different one, is still unanswered.

Finally, it should be mentioned that as early as in 1959, Hoffmman-Ostenhof proposed two hypotheses explaining the emergence of proteins with enzymatic properties (Hoffmann-Ostenhof 1959). First, polypeptides and proteins picked up metal ions or their complexes that were present in the primordial soup and used them as prosthetic group. The new combination yielded an enhanced or more specified catalytic activity. Second, proteins might possess by chance catalytic activities that were not the consequence of a log selection and not correlated with a catalytically active small molecule. One example are the enzymes of insects that were capable of destroying the insecticide DDT. In the long period of time, the evolution of insects has required, insects never had the opportunity to come into contact with a molecule such as DDT and to develop a defense system by natural selection. The second example concerns enzymes in the tissues of several animals that were capable of hydrolyzing local anesthetics such as procaine or analgesics such as dolantin (pethidine). These medicaments are synthetic chemicals which don't exist in nature. The meaning of these findings for the hypothesis of CE of proteins has apparently not been discussed yet. Concerning the aspect "unusual properties of proteins," an article of Lee et al. (1996) is worth mentioning which reports on self-reproducing oligopeptides.

A comment concerning the evolution of the protein synthesizing machinery in ribosomes is presented in Sect. 8.3.

8.2 Polynucleotides, Ribozymes, and the Evolution in the Test Tube

If we want to explain the prebiotic formation of the RNAs known from living organisms today, we have the same difficulties as for the explanation how proteins with enzymatic properties were formed. The problems arise immediately at the beginning of the synthetic pathway, namely with the question how all four nucleobases relevant for RNAs were formed simultaneously in similar quantities. Among the numerous model experiments compiled in Sect. 5.2, only one example with success in this direction can be found (Anders et al. 1974). However, the relevance of that experiment for the CE is doubtful and needs a critical discussion. Anders et al. used a kind of Fischer-Tropsch process with a mixture of CO, H_2, and NH_3 which requires high temperature and high pressure. As already discussed in Sect. 2.1 and in Chap. 4, it is absolutely unlikely that such reaction conditions existed on the early Earth.

High temperatures (600–1000 °C) exist in gases of volcanoes, but not the high pressures required for the Fischer-Tropsch process. Furthermore, NH_3 is absent from volcanic gases, and volcanic gases don't pass over Fe/Ni alloys which are needed as catalyst for this process. Moreover, it is hard to understand how a sudden cooling from 800–1000 °C down to temperatures <100 °C was possible under prebiotic conditions. A formation of nucleobases in the gas phase around 1000 °C must have followed by longer survival of hot and warm water until syntheses of nucleotides and RNA became possible in special compartments at low temperatures. However, cytosine is sensitive to hydrolysis and cannot have survived man thousands of years in hot or warm water. One more problem is high concentration of NH in the atmosphere, because this also means a considerable concentration in the primordial soup, where NH has prevented any synthesis of proteins or RNAs. This discussion touches a standard problem of most model experiments of prAAs and nucleobases or nucleotides. It does not help much to strengthen the hypothesis of CE when the conditions favoring the synthesis of one sort of biomonomers prevent syntheses of proteins and RNAs.

One more interesting and, unfortunately, negative results is the fact that cytosine was not detected in space or in meteorites. Even in the case of the more stable uracil, only traces were detected in one meteorite (Sect. 9.5). Hence, we have here an analogy to the prAA chemistry, because the 20 prAAs have never been formed in one model experiment and never been extracted from meteorites. Furthermore, there is the question to what extent non-RNA nucleobases and similar heterocycles (e.g., benzimidazole) were formed under prebiotic conditions and why the four RNA bases were selected. This problem is ignored ion almost all articles and books dealing with the emergence of the genetic code. Most researchers who performed model syntheses of nucleobases did not care about the question to what extent non-RNA nucleobases and other N-heterocycles were formed as by-products. Hayatsu et al. (1968, 1972) and Anders et al. (1974) reported that their analyses of reaction mixtures resulting from the Fischer-Tropsch process contained numerous N-heterocycles (s. Sect. 5.2). In this direction, only one study of meteorites has seemingly been published (s. Sect. 9.5) in which xanthine, hypoxanthine, and 2,6-diaminopurine were detected. All three nucleobases are capable of base pairing. How and why were the four RNA nucleobases selected? What were the criteria for their selection in an undirected evolution? Thermodynamic stability of the RNAs and their double helices cannot serve as argument for the following reasons:

1. Xanthine, hypoxanthine, and 2,6-diaminopurine possess nearly the same thermodynamic stability as adenine and guanine. Furthermore, methyl cytosine, dimethyl uracil, ammeline, melamine and cyanuric acid possess all the same thermodynamic stability as uracil and cytosine.
2. Most of the afore-mentioned heterocycles can undergo base pairing like the four RNA bases.
3. As demonstrated by the research group of Eschenmoser (1999, 2003), RNA analogs based on various pentoses or hexoses can form double helices of higher thermodynamic stability than the ribofuranose-based biological RNAs.

If the thermodynamic stability was not decisive for the selection of biological RNA what else was the criterium for their selection in anon-directed evolution? Of course, one may here again use the "a posteriori argument" that such a selection was unavoidable, because all living organisms use the same type of RNAs. However, when researcher adheres to this line of argumentation and when he adheres to the deterministic view, any research for reasons of the natural selection in an undirect evolution becomes obsolete.

When a plausible and experimentally tested route for the prebiotic formation of nucleoside triphosphates has been found (not definitely proven yet), and when their formation in compartments that hold back any disturbing chemicals (highly unlikely) can be demonstrated, this favorable situation allows one to design a plausible scenario for the selection and evolution of the biogenic RNAs. The beginning of research in this direction, currently performed by various research groups under the label "evolution in the test tube," began with experiments of S. Spiegelman and coworkers (Mills et al. 1967). The first step in those experiments was isolation of a fragment of the Oß-virus, which was mixed in a test tube with an enzyme isolated from the same virus and capable to catalyze the replication of RNA fragments. Those copies that showed the fastest replication were separated from the reaction mixture and transferred to another test tube where they were combined with a monomer mixture and with the replication enzyme. This process was repeated several times, and in this way, an isolation of the fastest growing RNA fragment was achieved which was then nick-named "Spiegelman's monster." However, this fragment had lost the capability to generate a complete virus.

As already mentioned in Sect. 2.3, the research groups of S. Altman and T. Czech discovered in the early 1980s where RNAs may have catalytic "talents," but the first examples were only capable of splitting RNAs (Guerrier-Takada et al. 1983; Czech 1990). T. Czech who together with S. Altman was awarded the Nobel Prize, called RNAs having catalytic properties "ribozymes." In the following decades, ribozymes played a decisive role in all research activities concerning "evolution in the test tube." For the sake of the historic order, the work of the Nobel Prize laureate M. Eigen should be mentioned first (Biebricher et al. 1981a, b, 1985, 1993). He had developed a new theory about the evolution of the genetic code (together with C. Biebricher, s. next section). He postulated that the first genes were rather short, because large genes have a high risk to be affected and deactivated by mutations (e.g., resulting from hydrolysis or radioactive radiation) or by incorrect copying. M. Eigen modified the Spiegelman experiment in such a way of combining the mixture of the four nucleotide triphosphates required for the synthesis of RNAs with the replication enzyme used by Spiegelman. However, he avoided the presence of RNA fragments, and thus, the presence of a matrix supporting the polycondensation of the monomers. Nonetheless, small oligonucleotides were formed an immediately thereafter they were copied, so that polynucleotides having DPs up to 100 repeat units were formed. Those artificial RNAs did of course, not resemble the Qß-virus, but an "ab initio" synthesis of RNAs was developed, although a biogenic enzyme was used.

Grounded on the experiments of Eigen and grounded on the discovery of ribozymes, a new research field entitled "evolution in the test tube" (ETT) was developed, which the goal to elaborate a synthetic strategy an suitable synthetic method to generate a fist primitive cell in a reactor, starting out from simple chemicals that were certainly available on the early Earth. The exterior of this first cell should consist of a lipid double layer, and it should contain self-replicating ribozymes. Furthermore, these ribozymes should be capable to organize a kind of metabolism including syntheses of proteins on the basis of the simple starting materials penetrating the lipid layer. Moreover, they should be capable to synthesize and select suitable amphiphilic lipids to enable a division and proliferation of the cell. This concept represents the highest level of the bottom-up strategy and closes the gap to the minimal cell concept (s. Luisi 2006) of the top-down approach. The following enumeration and discussion of experiments and results concerning the ETT do not claim to be exhaustive, in as much as it is difficult to define borderlines to closely related types of research activities.

The majority of the publications in his field and the most important results were contributed by the research groups of G. Joyce, J. W. Szostak, and D. B. Bartel, who had started his career in the group of Szostak. The main goal of many studies was the production of pool of RNAs made up by a broad variety of sequences and to select, optimize, and isolate ribozymes having special catalytic activities. The capability of self-replication was in the foreground of these studies, because self-replication is one of the most prominent and important properties of livening organisms. Joyce and Orgel (2006) called this goal "dream of biologists."The first studies aimed at selection and amplification of special ribozymes from pool of many (sometimes thousands of) different RNAs were launched around 1990. Seemingly, the first successful contribution to the ETT was published by Robertson and Joyce (1990). Those authors prepared a modified ribozyme by chain extension of a natural ribozyme which was isolated from an intron of a certain microbe. Using the four canonic nucleotides as monomers for the chain extension, a new ribozyme was prepared which served as template for the synthesis of the complementary DNA. The most active polymerase was selected from six different modified RNAs, and the complementary DNA was then transcribed into an RNA, whereupon an amplification by factor 100–1000 was achieved. A similar selection strategy was studied by Szostak and coworkers, but different methods were applied and the goal was different (s. below).

In the following three decades, numerous pools containing hundreds or thousands of RNAs with random sequences were prepared by enzymatic or combinatorial synthetic methods, and ribozymes having specific catalytic properties were selected. The first selection step was followed by an amplification step using an RNA polymerase (PCR). This reaction sequence was repeated three or four times, and ribozymes with high specificity and high efficiency were finally isolated. The complete process required hard work over a period of several weeks. Therefore, it was a significant progress, when Cox and coworkers published in 2001 an automated procedure, which shortened the process time to a period of 3 or 4 days (Cox and Ellington 2001; Cox et al. 2002).

Concerning the work of Joyce and coworkers, the following results should be highlighted. In the years 1999–2001, the group succeeded to isolate ribozymes that were bare of cytidine (Rogers and Joyce 1999) and finally even a ribozyme consisting of a sequence of only two different nucleotides was prepared (Reader and Joyce 2002). These important results suggest that in an early stage of the CE, the evolution of ribozymes began with sequences of only two nucleotides which, by natural selection, broadened the spectrum of catalytic activities by incorporation of one and finally two more different nucleotides. Furthermore, the Joyce group developed enzymes that were active at relatively extreme pH values, for example, at pH 5.8 or pH = 9.8 (Kuehne and Joyce). These results are of interest with respect to the existence (and perhaps early evolution) of microorganisms such as archaea that can exist at such extreme pHs. Moreover. The in vitro transformation of a ribozyme into a deoxy-ribozyme is worth mentioning (Paul et al. 2006) and finally, the complexation of a D-RNA with an L-aptamer (Sczepanski and Joyce 2013; for the definition and importance of aptamers, seethe comments on the work of J. Szostak below).

An outstanding result of the research group of G. Joyce was the selection and amplification of ribozymes that were capable of self-reproduction. Those ribozymes were exclusively based on synthetic starting materials (NAs) (Voytek and Joyce 2007; Lincoln and Joyce 2009). When sufficient concentrations of all four canonic nucleotides were present in the reaction mixture, an exponential proliferation of the self-replicating ribozymes was observed.

Szostak and coworkers used the afore-mentioned in vitro selection method above all for the isolation of RNAs suitable as highly specific aptamers. At the same time, the "Gold-Laboratories" published the development of a similar procedure, called SELEX (Systematic Evolution of Ligands by Exponential Enrichment) designed to find aptamers for T4DNA-polymerase. The existence of aptamers specified on individual prAAs and nucleotides and nucleotide triphosphates is extremely important for the understanding of the evolution of protein syntheses in ribosomes (s. next Section). Szostak and coworkers initially focused their studies on isolation and characterization of aptamers specific for nucleoside mono- and triphosphates: Famulok and Szostak (1992a, b), Sassanfar and Szostak (1993), Lorsch and Szostak (1994), Hutzinga and Szostal (1995), Lauhon and Szostak (1995), Diekman et al. (1997), Holeman et al. (1998), Vaish et al. (2000, 2003), Davis and Szostak (2002), Huang and Szostak (2003) Sazani et al. (2004), Plummer et al. (2005), Carothers et al. (2006). All these positive results have, of course much contributed to the acceptance of the RNA world hypothesis, but the questions raised in Sects. 4.3, 11.1, and 11.2 are not answered yet.

After D. P. Bartel had left the Szostak group in the early 1990s, he formed his own research group, which elaborated a number of interesting and important contributions to the ETT concept: Ekland et al. (1995), Sabeti et al. (1997), Tuschl et al. (1998), Umrau and Bartel (1998, 2003), Johnston et al. (2001), Glasner et al. (2000, 2002), Merryman et al. (2002), Backerville and Bartel (2002), Müller and Bartel (2003, 2008), Bergman et al. 2004) Lawrence and Bartel (2003, 2005), Curtis and Bartel (2005, 2013), Schultes et al. (2005), Stechner et al. (2009), Stechner and Bartel (2011), Bagby et al. (2009). The most intensively studied working fields were in vitro selection and characterization of ribozymes, on the one hand, and primer-initiated oligomerization of nucleoside triphosphates catalyzed by ribozymes, on the other also, worth mentioning in the synthesis of pyrimidine nucleotides by means of ribozymes (Umrau and Bartel (1998). P. Umrau, who founded an own research group after having left the team of D. Bartel group and contributed the following publications to the ETT: Zaher and Umrau (2006), Zaher et al. (2006), Lau et al. (2004), Lau and Umrau (2009), Cheng and Umrau (2010).

The works of the British biologist p. Holliger are not directly concern with the ETT concept, but they concern with the RNA world/compartment hypothesis. In this connection, the following publications are of interest: Ghadessy and Hollinge (2007), Wochner et al. (2011), Pinheiro and Holliger (2012), Mutschler and Holliger (2014), Taylor et al. (2015), and Houlihan et al. (2017). In the first paper mentioned above, a new method is described called "compartmentalized self replication" (CSR).

This method is based on stable oil in water emulsions as compartments. In each oil droplet, a polymerase produces a gene which carries the genetic information for the biosynthesis of the polymerase. Another working field is the systematic variation of RNA and DNA sequences with the goal to find out, first, to what extent base pairing with natural RNAs or DNAs is possible, and second, if reproduction by polymerase is feasible, so that they may be subjected to a Darwinian evolution with optimization of certain properties. This research direction may help to clarify, why the RNAs and DNAs extant in living organisms were selected in the course of CE. Holliger and coworker varied at first the saccharide unit, and the resulting nucleotide analogs were labeled XNA. Other variations of the RNA structure were performed by other authors (Nielsen 1995; Eschenmoser 1999; Loakes and Holliger 2009; Herdewijn 2010; Campbell and Wengel 2011). In 2012, Pinheiro and Holliger draw the following conclusion (Pinheiro and Holliger 2012):

> We also select XNA aptamers, which bind their targets with high affinity and specificity, demonstrating that beyond heredity, specific XNAs have the capacityfor Darwinian evolution and folding into defined structures. This heredity and evolution, two hallmarks of life, are not limited to RNA and DNA but are likely to be emergent properties of polymers capable of information storage.

His conclusion aggravates a decisive problem which is underlaying the entire concept of CE on the early Earth. Why did the selection of biomonomers and of biopolymers, mainly RNAs, take a course in direction of an autopoietic and self-replicating cell? All the numerous afore-mentioned successful experiments explain how individual reactions and reaction sequences in a CE might have occurred, but they don't explain why they have occurred. In the ETT research, it is the experimenter, who decides what kind of reaction products should be selected. But, who has told the numerous molecules involved in the CE that their selection should take a course which ended with the first living cell? More precisely, why were ribozymes selected that developed a machinery capable of synthesizing proteins? Why did certain ribozymes not collaborate to synthesize polyamides via connection of the ω-function of trifunctional amino acids such as Asp, Glu, Orn, Lys, and Arg? The ω-amino groups are slightly more reactive (nucleophilic) than the α-amino groups, and the resulting polyamides are thermodynamically stable and even less sensitive to hydrolysis than polypeptides. The structure of such polyamides is, of course, less favorable for adopting and exerting a broad variety of biological functions, but who has told the ribozymes what kind of biological functions are needed, in as much, as ribozymes can be efficient catalysts themselves? In other words, an undirected evolution of molecules takes a course towards increasing diversity of molecules under the dictate of entropy as evidenced by the space chemistry. A CE proceeding with a permanent selection of molecules in a certain direction is a course which involves reduction of entropy. The final goal, a living (1993), cell, is an object which is permanently engaged in a fight against entropy (e.g., racemization and mutation) by absorption and use of sunlight and by exothermic reactions such as oxidation of nutrients. All the results of model reactions including ETT don't explain why the CE has followed a steady course against entropy.

A detailed discussion of all the numerous research articles published by G.F.Joyce and coworkers or by J. W. Jostak and coworkerds is beyond the scope of this book. Therefore, a list of the articles not cited above follows here (in alphabetic order of the first author). Joyce group: Beaudry and Joyce (1993), Breaker and Joyce (1994a and b), (1995), Dai and Joyce (2000), Dai et al. (1995), Horning and Joyce (2016), Jaeger et al. (1999), Joyce (1989), (1992), (2002b), (2007a and b), Kim and Joyce (2004), Koehne and joyce (2003), Kuhns and Joyce (2003a and b), Kumar and Joyce (2003), Lam and Joyce (2009), (2011), Lehman and Joyce (1993a and b), McGinnes and Joyce (2002), McGinnes et al.(2002), Olea et al. (2012), Ordoukhanian and Joyce (2002), Pagel and Joyce (2010), Petrie and Joyce (2014), Rogers and Joyce (2001), Robertson and Joyce (2014), Santoro and Joyce (1997), (1998), Sczepanski and Joyce (2012), (2014), Sheppard et al. (2000), Tsang and Joyce (1994), (1996), Szostak group: Bartel and Szostak (1993), Blain and Szostak (2014), Budin and Szostak (2010), Campbell and Szostak (2006), DeRicardo and Szostak (2009), Doudna et al. (1993), Green and Szostak (1991), Hager and Szostak (1997), Lee et al. (2000), Lohse and Szostak (1996), Luptak and Szostak (2008), Powner et al. (2010), Saleh-Ashitani and Szostak (1999), Suga and Szostak (1998), Wilson and Szostak (1992), (1995), Woriskay et al. (2011). Finally, review articles concerning ETT and the RNA world hypothesis should be cited (in alphabetic order of the names of the research group leaders):

A. D. P. Bartel: Bartel (1999), Bartel and Umrau (1999).
B. P. Holliger: Pinheiro and Holliger (2012, 2014), Pinheiro et al. (2013a, b), Attwater et al. (2013) Attwater and Holliger 2014), Taylor et al. (2015), Houlihan et al. (2017), Pinheiro et al (2012).
C. G. F. Joyce: McGinnes and Joyce (2003), Paul and Joyce (2004), Joyce (2000, 2002a, b, 2004a, b, 2012), Joyce and Orgel (2006), Robertson and Joyce (2012), Breaker and Joyce (2014), Olea and Joyce (2016), Breaker and Joyce (1995), Breaker et al. (1994a, b), Olea et al (2015), Santoro and Joyce (1998), Santoro et al. (2000).
D. J. W. Szostak: Szostak (1997, 2011, 2017), Famulok and Szostak (1992a), Wilson and Szostak (1999), Saleh-Asdhirani and Szostak (2001), Chen et al. (2006)Carothersetal.(2006)Trevinoetal.(2011),ChapmanandSzostak(1995a,b).
E. P. J. Umrau: Wang et al. (2010).

8.3 The Genetic Code

The experiments discussed above lend strong support to the RNA world hypothesis but they don't explain yet, how the genetic code was developed which had the consequence that numerous different proteins with a broad variety of catalytic activities could and can be synthesized in a controlled and reproducible way. The genetic code consists of a sequence called codons which are equivalent of the individual prAAs. Each codon consists of a sequence of three nucleotides (s. Table 8.1). Since the structure of RNAs is based on four different nucleotides (A, C, G, U, Fig. 8.1), their combination yields 64 different codons. Since only 20 different prAAs coding, several prAAs are coded by two, three, or even four different codons (Table 8.1). These different codons may have the purpose to indicate the position of a prAA inside the protein chain, for example, at the beginning or at the end of protein chain.

After the discovery of the double-helix structure (s. Fig. 8.2b) of the DNA by the Nobel Price laureates F. Crick and J. Watson in the year 1953, numerous research groups began with studies aiming at the decipherment of the genetic code (Fig. 8.2). In March 1961, H. Matthei published the first identification of a codon, namely UUU as codon for Phe. In the following 5 years, most other codons were identified, and when in 1966, a conference entitled "the genetic code" was organized in Cold Spring

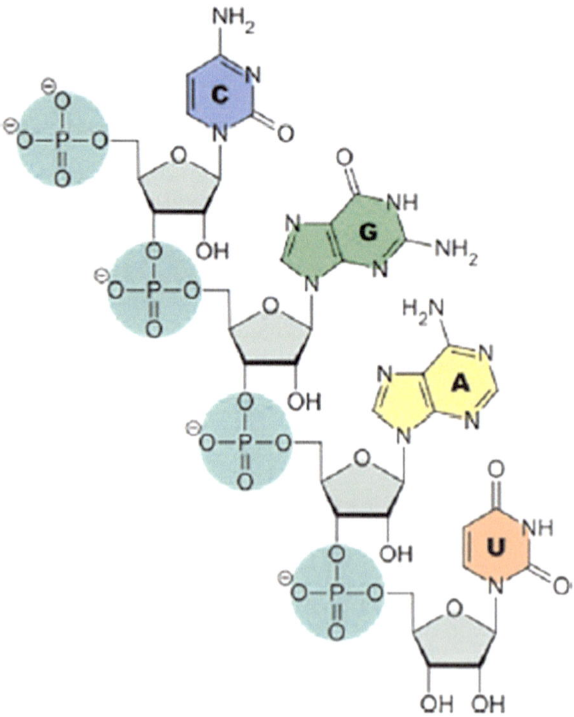

Fig. 8.1 Segment of a RNA sequence

Table 8.1 prAA—codon correlation ordered according to increasing number of available codons for one prAAs[a]

prAA	Codon	prAA	Codon
Initiator	AUG	His	CAU, CAC
Met	AUG	Lys	AAA, AAG
Trp	UGC	Ile	AUU, AUC, AUAGGU, GGC; GGA, GGG
Sec	UGA	Gly	GCU, GCC, GCA, GCG
Tyr	UAU, UAC	Ala	GUU, GUC; GUA; GUG
Cys	UUU, UUC	Val	ACU, ACC, ACA, ACG
Asn	UGU; UGC	Thr	CCU, CCC, CCA, CCG
Asp	GAU, GAC	Pro	CUU, CUC, CUA, CUG, UUA; UUG
Gln	CAA; CAG	Leu	UCU, UCC, UCA, UCG, UUA, UUG
Glu	GAA, GAC	Ser	CGU, CGC, CGA, CGG, AGA, AGG
		Arg	

[a]There is no codon for 4-hyroxyproline, because this prAA is formed by oxidation of Pro in collagen

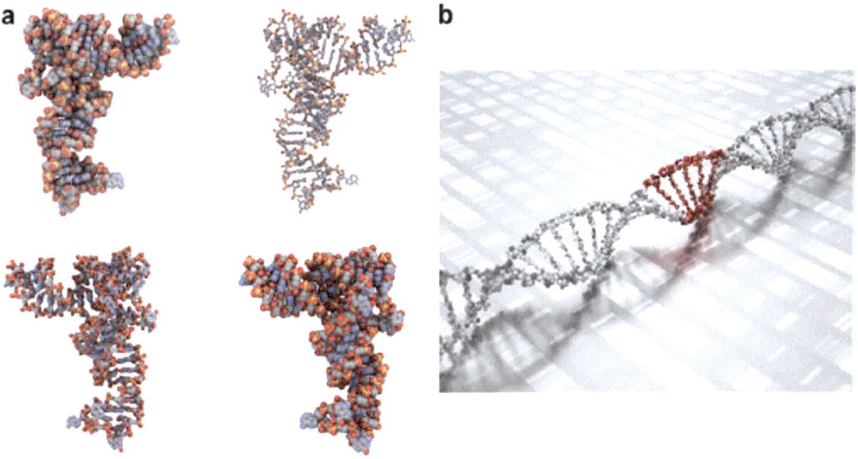

Fig. 8.2 (**a**) 3D Structure of a transfer RNA containing double-helix segments (© molecule stock adobe com), (**b**) schematic illustration of a segment of a double helix characteristic for DNA (© sunnyboy 92/Fotolia)

Harbor, almost all codons were deciphered. In the following year, a book with the same title was published by the biologist Woese (1967). One year later, Crick (1968) and Orgel (1968) first proposed the hypotheses about the origin of the genetic code.

The salient points of Crick's hypothesis may be summarized as follows. He assumed that transfer (t) RNA and ribosomal (r) RNA were from the very beginning of the fundamental components of a protein synthesizing machinery. The modern code emerged step-by-step out of a primitive origin. At first, the four canonical nucleotides served as codons for only four prAAs and somehow arranged the synthesis of copolypeptides exclusively consisting of these four prAAs. Since these copolypeptides were not well suited as carriers of various different catalytic properties, the selection improved this system by inventing and favoring RNAs consisting of dual codons made up by two nucleotides. In this way, RNAs consisting of 16 different codons evolved capable to code for 16 different prAAs. Finally, the evolution decided that the "modern" version of RNAs containing codons of three nucleotides was more flexible and more suited for a production of a large variety of proteins serving numerous different biological functions.

In his article of 1968, L. E. Orgel, who was in permanent contact with F. Crick, presented a view which was rather complementary than alternative to the concept of Crick. He compared the "Protein-World" with the "RNA-World" hypothesis and provided little new insights into the evolutionary process that yielded the genetic code. He emphasized that the interaction between prAAs, RNAs, and proteins was based on the normal laws of nature known in chemistry and physics:

> While great deal remains to be done, experiments on prebiotic syntheses suggest that all of the steps required for polynucleotide and polypeptide formation could have occurred, albeit inefficiently, in the absence of biological catalysts.

This optimistic view is, of course, understandable for the state of knowledge in the year 1968; Orgel continued:

> It is important in arguments about biochemical evolution to distinguish between two stages in the development of the process of natural selection. In the first stage suppose the components for building the polymers were freely available the environment and that the natural selection favored the selection of those sequences best able to replicate. I call this "natural selection without function". It seems reasonable to require that the process proposed in "natural selection without function" must be understandable rather directly in terms of the physical chemistry of the elementary components. All events must be probable in the light of the ordinary physico-chemical interactions between the components of the system. Such processes should be demonstrable in the laboratory in experiments occupying a reasonable short time.

In the year 1981, the biophysicists H. Kuhn and J. Waser published a treatise on "Molecular self-organization and the origin of Life" in which they describe their views as follows (translated by the author) (1981):

> We consider a series of numerous physical and chemical model reactions resulting in self-organization of matter. The progress of this process is driven by changing temperatures and by a broad variation of the environment, meaning different environmental conditions as they have existed on a prebiotic planet at least in several places. Such a special theoretical approach defines the frame in which the process of self-organization has occurred. One finds that in several barriers need to be cleared, part of which results from an accumulation of miscopies. An earlier barrier will be surpassed in such a way that certain miscopies will be filtered out by aggregation. Another barrier will be cleared by the evolution of a chemical machinery capable of producing a cellular membrane which concentrates all the important components inside. In this way a system evolves which produces a primitive replicase, a process stabilized by a primitive genetic code. A later barrier will be surpassed by a transformation and diversification of the first machinery in separate aggregates specialized on replication and translation of the genetic code.[....] This approach yields information about prerequisites, logic structure and organization of evolutionary process.

These comments are rather trivial, since evolution of a genetic code with production of proteins will necessarily proceed in a stepwise manner, otherwise the process does not obey the definition of evolution. The entire concept represents a kind of "a posteriori" argumentation, and any experimental evidence for the individual steps of this process is missing. It fits in with this superficial nature of Kuhn'and Waser' concept that those authors assume that all the molecules needed for the evolution of genetic code were available in the primordial soup. The literature cited in this connection is so superficial that it pretends to provide experimental facts that don't exist. For example, those authors say "that ribose was formed by condensation of formaldehyde in the presence of aluminum oxide and kaolinite", and they cite a paper of Gabel (1967). Yet, that paper does not contain any data about the synthesis of ribose. Furthermore, the formation of adenine from hydrogen cyanide (s. Sect. 5.2) is mentioned as evidence for the early availability of the canonic nucleobases, but those authors don't mention that under the conditions yielding adenine, pyrimidine bases cannot be synthesized. As complementary evidence Kuhn and Waser cite the synthesis of cytosine from cyanoacetylene, but they ignore that under those conditions, purine bases are not accessible.

Since 1971, the biophysicist M. Eigen developed a theory explaining the evolution of the genetic code and of the biosynthesis of proteins via molecular selection and self-organization of RNAs and proteins (Eigen 1971, Eigen and Schuster 1977, 1978, Eigen and Winkler-Oswatisch 1981, Eigen et al. 1981). His theory started out from the consideration that formation of proteins and high molar mass RNAs by chance via random reactions of biomonomers in the primordial soup was absolutely unlikely and did not justify scientific studies. His calculation of probabilities (1971, p. 469) yielded similar results as the calculation published by F. Hoyle, H. Morowitz, and R. Shapiro (Sect. 2.1). Eigen assumed that as soon as small polypeptides and short RNA sequences appeared on the early Earth, a selection process set out analogous to the Darwinian selection of living organisms, which favored the formation of proteins and RNA sequences similar to those existing in living organisms today. With L. E. Orgel, H. Kuhn, and I. Waser, he shared the overoptimistic belief that all the components required for a successful CE were available in the primordial soup (1971):

> All we need at the moment is to assume that substances like energy rich phosphates, activated amino acids etc. were present and could be condensed into macromolecular substances exhibiting single catalytic functions.

This assumption which is not only overoptimistic but even wrong, considering the facts known in the early twenty-first century was already criticized by Shapiro who cited Eigen's view as follows (1987, p. 171):

> Here we simply start from the assumption that when self-organization began, all kinds of energy rich materials ere ubiquitous including in particular amino acids in degrees of abundances, nucleotides involving the four bases A, U, G, C, polypeptides of both preceding classes [.....] having more or less random sequences.

Concerning the selection mechanism, Eigen said (1971):

> If we want to close the gap between physics and biology, we have to find out what selection means in precise molecular terms which can ultimately be described by quantum-

> mechanical theory. We have to derive Darwin's principle from the known properties of matter.

He also believes that the interactions between functional groups (in the side chains) of the copolypeptides, on the one hand, and the base pairing of nucleobases, on the other hand, play an important role. These interactions and catalytic activities which attributes to even simple copolypetides having nearly random sequences are important components for a successful selection process (1971, p. 471):

> Catalytic function in combination with various fed-back mechanisms causing certain self-enhancing growth properties of the system, will be shown to be one of the decisive prerequisites for self-organization.

An important aspect addressed by several scientist in their hypotheses about the emergence of the genetic code and its capability of self-reproduction is was the question, how the evolution was able to optimize the genetic code, so that its evolution survived the occurrence of numerous defects and still survives the permanent occurrence of various defects. Defects may result from events such as incorrect copying of a RNA or DNA sequence and chemical damages by radicals, by UV light or by radioactive radiation. It was postulated, not only by M. Eigen, that in the beginning of the evolution of the genetic code, there was an optimum length of the ribozymes to enable an successful selection process because oligomers were too short for storage of sufficient information, whereas extremely long chains were too sensitive to various damages. This consideration prompted to questions. First, how long was the optimum length of the ribozymes, and second, it was possible that such RNAS could be formed in the primordial soup or suitable compartments. M. Eigen, who was the first to ask these questions, draw the conclusion that the optimum replicator should have had a length of approximately 100 nucleotides, but that the formation of such RNAs in a primordial soup was absolutely unlikely. As solution of this problem, he and the biologist Schuster designed the hypothesis of a "hyper-cycle" (Eigen and Schuster 1977, 1978). Those authors explained the nature of a "hypercycle" as a combination of several complementary smaller reaction cycles involving and optimizing catalytic activities (1977):

> Consider a sequence of reactions in which at each step, the products with or without the help of additional reactions, undergo further transactions. If in such a sequence any product formed is identical with a reactant of a proceeding step, the system resembles a reaction cycle, and the cycle as whole a catalyst.

This means for the formation of optimum replicators that at first a short RNA was formed (RS1) which favored the formation of another RNA (RS2) which catalyzed, in turn, the formation of RS3 until a RSx emerged which catalyzed the formation of RS1. In combination with selection mechanism such a "hypercycle" should finally result in the formation of large RNA replicators capable of efficient storage of information. This concept looks pretty hypothetical and optimistic, and ERigen and Schuster exclusively presented a mathematical treatment without any experimental background. However, this concept is in fact realized in numerous metabolic cycles of living organisms. Considering several more or less plausible assumptions Eigen and Schuster reached the following final conclusion (1977):

> The results of our studies suggest that the Darwinian evolution of species was preceded an analogous stepwise process of molecular evolution leading to an unique cell machinery which uses an universal code. This code came finally established, not because it was the only alternative, but rather due to peculiar "once-forever" selection mechanism which could start from any random assignment. Once-forever is a consequence of hypercyclic organization. A detailed analysis of macromolecular reproduction mechanisms suggests that catalytic hypercycles are a minimum requirement for a macromolecular organization that is capable to accumulate, preserve and process genetic information.

The hypotheses and calculations of Eigen and Schuster have influenced and stimulated other research groups, but they also have attracted punctual critique. For example, G. Witzany (2016) has criticized the reductionistic, physicochemical view of Eigen as too one-sided:

> The genetic code with its typical language-like features (characters adopt syntactic semantic pragmatic rules) take the stage of a real natural language with interactional group-building of various RNA stem-loops. The interactional and group-building cooperativity of the RNA stem loops constitute the genetic code as a real natural code, not its physicochemical key characters alone.

Considering the modern understanding of the ribosomal protein synthesis, Eigen's interpretation of prAA-specific aptamers as early precursors of transfer RNAs plays an important role (Eigen and Winkler-Oswatitsch 1981). At first, short RNAs were formed capable of forming large loops that were capable to undergo specific interactions with individual prAAs (Fig. 8.2a). Under the catalytic influence of ribozymes, the carboxyl groups of the prAAs were activated by means of ATP as mixes anhydrides and transferred to the 3′-OH group of a suitable tRNA. Since for the 20 prAAs twenty specific tRNAs are required, one may attribute an aptamer character to the tRNAs, although the connection of prAAs to tRNAs is a covalent bond in contrast to the association typical for AA-aptamer conjugates. The tRNA carries the prAAs to that site of the ribosome, where the messenger (m) RNA dictates the sequence of their polymerization.

Concerning the stepwise course of the pRAA-tRNA esterification and the formation of proteins, Szathmary and Maynard-Smith proposed the following scenario (Szathmary 1993, 1999, Maynard-Smith and Szathmary 1993, Szathmary and Maynard-Smith 1997):

A. A decisive driving force for the esterification of pRAAs with tRNAs was at first not the production of proteins, but an enhancement and diversification of catalytic activities of the ribozymes.
B. The prAA-oligonucleotide conjugate, precursor of the tRNA, underwent base pairing with a longer, loop forming RNA and modified its catalytic property.
C. Ribozymes catalyze the formation of prAA-oligonucleotide complex in such a way that (optimized by selection) specific prAA nucleotide sequences are favored over longer period of times.

D. prAAs are preferred as cofactors of ribozymes, because Miller's experiments have shown that prAAs were available in the primordial soup (at this point not only wishful thinking, but also a lack of knowledge about the Miller experiments come into play).
E. The ribozymes with modified catalytic activities were subject to evolutionary process which over longer periods of time resulted in protein enzymes.

Concerning the correlation of individual prAAs with specific codons, those authors wrote (translated from German by the author):

Although a couple of properties of the genetic code may reflect chance events, there are other properties that resulted from optimization processes and suggest that the correlation was a consequence of selection.

1. Similar codons are specific for the same prAA, zB GUU, GUC, GUA, and GUG for Val.
2. prAAs with similar properties correspond to similar codons, for example, GAU and GAC for Asp and GAA and GAG for Glu.
3. prAAs which were high abundance in proteins are represented by a relatively large number of slightly different codons, for example, four codons are active for the frequently occurring Leu, whereas the rare AA Trp is only represented by one codon.

The first two correlations have and had the purpose to minimize negative effects of mutations.

In this context, a complementary comment of Freeland and Hurst (1998) should be cited:

> Statistical and biochemical studies of the genetic code found evidence of non-random patterns in the distribution of codon assignments. It has, for example, been shown that the code minimizes the effect of point mutations ad mistranslations: erroneous codons are either synonymous or code for an amino acid with properties very similar to those of the one that would have been presented had the error not occurred. This work has suggested that the second base of the codons is less efficient in this respect, by about three orders of magnitude than the first and third one.

At the end of their summary:

> We thus conclude not only that the natural genetic code is extremely efficient in minimizing the effect of errors, but also that its structure reflects biases in those errors, as might be expected where the code is the product of selection.

This stance has found broad acceptance over the past two decades. The genetic code of living organisms represents, based on a long selection process, an optimization of its stability with respect to damages resulting from mistranslations, point mutation, and chemical accidents such as racemization of prAAs. However, the facts that the genetic code has been deciphered and the reaction pathway of the protein synthesis has been elucidated do not answer the following questions:

1. How and why did the evolution develop aptamers for the 20 prAAs and not for the numerous other non-protein AAs or imino acids, although many of them

were more abundant than several prAAs? It is absolutely unlikely and not prove that compartments existed on the early Earth equipped with membranes capable of separating prAAs from non-protein AAs, in as much as the properties of several prAAs were quite similar to those of non-protein AAS (e.g., Ala versus α-Abu). Furthermore, it is quite unlikely an unproven that among the large number of RNA sequence that may have existed in an early stage of the CE, aptamers suited for binding non-protein AAS were absent.
2. Even when prAA-aptamer complexes were preferentially formed, how and why was a genetic code formed favoring syntheses of proteins that were efficient catalysts, since the ribozymes were themselves good catalysts with a broad variety of catalytic activities.

The latter question, already mentioned at the end of the preceding Section, should be discussed here in more detail. The development of a machinery capable of synthesizing numerous special proteins and the continuous operation of this machinery require a lot of energy and involve a permanent fight against entropy. The chemical evolution of the space chemistry demonstrates what happens under the dictate of entropy, namely a steadily increasing diversity of molecules and not an increase of molecular order and specialization on certain complex biopolymers. Therefore, a CE on Earth proceeding from gases such as CO_2, N_2, and H_2O to a living cell is far from being a normal consequence of the normal laws of chemistry and physics and requires an explanation how and why it was directed toward a first cell.

Consider as first example, the discovery and elaboration of ribozymes are capable of self-replication, as reported by G Joyce and coworkers. What does this important scientific success mean as argument for a successful CE? The positive aspect is, of course, that self-replicating ribozymes may be understood as a first primitive model of a self-replicating cell. Now it has to be considered that RNAs in warm aqueous solution are subject to a steady hydrolysis, in as much, when the pH deviates from 7.0. A self-replicating ribozyme will permanently find new monomers for the continuation of its self-reproduction resulting from the degradation of other RNAs. Hence, the population of the self-replicating ribozyme (Rz1) will steadily grow at the expense of other RNAs. The population of Rz1 will be subjected to point mutations and miscopying, and variants will be formed that are inactive or less active. However, by chance, a new variant may appear (Rz2), which is capable of a more efficient self-replication than the original Rz1 with the consequence that Rz2 will more or less replace Rz1 with increasing time. This process may happen again and a selection of self-replicating ribozymes toward the fittest for survival and growth will take place (RzX), a perfect Darwinian scenario.

RzX corresponds on the level of chemical evolution to egoistic gene proposed by the biologist R. Dawkins as goal and driving force in the biological evolution. However, biologist have correctly criticized Dawkins concept, because a living cell needs cooperation of all genes for its survival and proliferation (Bauer 2008). Egoism of individual genes is deadly for any cell, and an evolution of cells on the basis of egoistic genes is a contradiction in itself. This line of argumentation is also valid for self-replicating enzymes. The formation of an RNA machinery capable of synthesizing proteins requires a precise cooperation of numerous different RNAs, and an egoistic ribozyme capable of efficient self-replication would prevent the development of protein synthesizing ribosomes. At this point, the question arises, how did the ribozymes learn to cooperate for the purpose of protein syntheses and to avoid egoistic self-reproduction.

Consider as second example the maximum aim of (bio)chemists working on the ETT concept, namely the synthesis of a self-reproducing living cell exclusively based on synthetic components, above all on synthetic ribozymes. The realization of this aim would be an extraordinary scientific success and it would strongly support the RNA world hypothesis. Yet even this success would not give an answer on the question, why a CE must have taken a course ending with the formation of a living cell. In the case of the ETT, all experiments are performed and controlled by researchers who have the blueprint of a living cell in mind and design and optimize the synthetic strategy in a way which favors a successful course. The researchers know what kind of reaction products and intermediates are useful to reach the final goal and which intermediates and by-products should be removed. However, who has informed the molecules produced in the CE about the blueprint?

The "a posteriori" argument, that the molecules must have known what they should do because otherwise the first cell would not have been "invented" and the humankind would not exist and study the hypothesis of CE is not a serious scientific answer. Since the reductionistic-deterministic concept must be abandoned, the "a posteriori" argument says that all the matter involved in the CE must have known what to do to realize a successful CE. At this point the aposteriori" argumentation joins the view of the "anti-reductionists" who attribute live an added value" which goes beyond a simple combination of chemical and physical laws of nature. In other words, the design of the universe includes a law of nature which complementary to the normal laws of chemistry and physics allows for a successful CE toward a living cell, whenever the reaction conditions are suited for such a process. The ongoing examination of exoplanets may give an answer on his question, but only when the answer is positive. Even when hundreds of exoplanets were studied, a negative result does mean much, because the timely and spatial distance between exoplanets and events producing a living cell might be so huge that the humankind and other intelligent organisms don't have a chance to discover each other.

Finally, a couple of review articles and book concerning the evolution of the genetic code that have not been mentioned above should be cited: Vaas (1994), Kay (2000), Zubay (2000), Ribas de Pouplan (2004), Wang and Schultz (2005).

References

Anders E, Hayatsu R, Studier M (1974) Orig Life 5:57

Attwater J, Tagami S, Kimoto M, Butler K, Kool E, Wegel J, Herdeweijn P, Hitao T, Holliger P (2013) Chem Soc 4:2804

Attwater J, Holliger P (2014) Nat Methods 11:495

Bagby S, Bergman N, Stechner D, Yen C, Bartel D (2009) RNA 15:2129

Bartel D (1999) Cold Spring Harb Monogr Arch 37:143 (The RNA world. 2nd edn)

Bartel D, Szostak J (1993) Science 261:1411

Bartel D, Umrau P (1999) Trends Cell Biol 9:M9–M13 and Trends Biochem Sci 24:M9–M13.

Backerville S, Bartel D (2002) Proc Natl Acad Sci USA 99:9154
Bauer J (2008) Das kooperative Gen. Hoffmann und Campe, Hamburg
Beaudry A, Joyce G (1993) Science 257:635
Bergman N, Lau N, Lehnert H, Westhof E, Barttel D (2004) RNA 10:176
Biebricher C, Eigen M, Luce R (1981a) J Mol Biol 148(4):369
Biebricher C, Eigen M, Luce R (1981b) J Mol Biol 148(4):391
Biebricher C, Eigen M, Gardiner Jr W (1985) Biochemsitry 24(23):n6550
Biebricher C, Eigen M, McCatskill J (1993) J Mol Biol 231:175
Billmeyer F (1984) Textbook of polymer science.3rd edn. Wiley, New York
Blain J, Scpostak J (2014) Annu Rev Biochem 83:6215
Breaker R, Joyce G (1994a) Proc Natl Acad Sci USA 91:6093
Breaker R, Joyce G (1994b) Chem Biol 1:223
Braeker R, Joyce G (1995) J Mol Evol 40:551
Breaker R, Joyce G (2014) Chem Biol 21:1059
Beaker R, Barnerji A, Joyce G (1994) Biochemistry 33:11980
Budin I, Szostak J (2010) Annu Rev Biophys 39:245
Campbell M, Wengel J (2011) Chem Soc Rev 40:5680
Campbell M, Sziostak J (2006) Aptamer handbook. pp 3–28
Carothers J, Östereich S, Szostak J (2006) J Am Chem Soc 129:7929
Cedrangolo F (1959) The problem of the origin of the proteins. In: Clark F, Synge R (eds) The origin of life on the Earth. Pergamon Press, London
Chapman K, Szostak J (1995a) Curr Opin Struct Biol 4:618
Chapman K, Szostak J (1995b) Chem Biol 2:325
Cheng L, Umrau P (2010) CSH Perspect Biol 2(10):a002204
Chen L, Hanczy M, Sazani P, Szostak J (2006) Cold Spring Harb Monogr Arch 43:57 (RNA world. 3rd edn)
Cox J, Ellington A (2001) Bioorg Med Chem 9:2525
Cox J, Hayhurst A, Hesselberth J, Bayer T, Georgiu G, Ellinton A (2002) Nucl Acid Res 30:e108
Crick F (1968) J Mol Biol 38:367
Curtis E, Bartel D (2005) RNA 11:1173
Curtis E, Bartel D (2013) RNA 19:1113
Czech T (1990) Biosci Rep 10:239
Dai X, Joyce G (2000) Helv Chim Acta 83:1701
Dai X, De Mesmacker A, Joyce G (1995) Science 267:237
Davis J, Szostak W (2002) Proc Natl Acad Sci USA 99:11616
DeDuve C (2002) Life evolving: molecules, mind and meaning. Oxford University Press, Oxford
DeRicardo A, Szostak J (2009) Sci Am 301:54
Diekman T, Butcher S, Sassanfar M, Szostak J (1997) J Mol Biol 272:467
Doudna J, Usman N, Szostak J (1993) Biochemistry 32:2111
Eigen M (1971) Narurwissenschaften 58:465
Eigen M, Schuster P (1977) Naturwissenschaften 58:465
Eigen M, Schuster P (1978) Naturwissenschaften 65:341
Eigen M, Winkler-Oswatitsch R (1981) Naturwissenschaften 68:217
Eigen M, Gardiner W, Schuster P, Winkler-Oswatitsch R (1981) Sci Am 144(4):88
Ekland E, Szostak J, Bartel D (1995) Science 269:364
Elias H (1997) An introduction to polymer science. Wiley VCH, Weinheim
Eschenmoser A (1999) Science 284:2118
Eschenmoser A (2003) Creating a perspective for competing. In: Proceedings of the Templeton Foundation "Biochemistry and fine tuning". Harvard University, pp 10–12
Famulok M, Szostak J (1992a) J Am Chem Soc 114:3990
Famulok M, Szostak J (1992b) Angew Chem Int Ed 31:979
Freeland S, Hurst L (1998) J Mol Evol 47:238
Gabel N, Ponamperuma C (1967) Nature 216:453
Ghadessy F, Hollinge P (2007) Methods Mol Biol 352:237
Glasner M, Yen C, Ekland E, Bartel D (2000) Biochemistry 39:15556

Glasner M, Bergmann N, Bartel D (2002) Biochemistry 41:8103
Green R, Szostak J (1992) Science 258:1910
Green R, Ellington A, Bartel D, Szostak J (1991) Methods (SanDiego) 2(1):75
Guerrier-Takada C et al (1983) Cell 35:849
Hager A, Szostak J (1997) Chem Biol 4:607
Haldane J (1929) The origin of life. Ration Annu 148:3 (Reprint in Bernal 1967, pp 242–249)
Hayatsu R, Anders E, Oda A, Fuse K, Anders E (1968) Geochim Geophys Acta 32:175
Hayatsu R, Studier M, Matsuoka S, Anders E (1972) Geochim Geophys Acta 36:555
Herdewijn P (2010) Chem Biodivers 7:1
Hoffmann-Ostenhof O (1959) Der Urspung der Enzyme. In: Clark M, Synge K (eds). Pergamon Press London, The origin of life on earth
Holeman L, Robinson S, Szostak J, Wilson C (1998) Fold Des 3:423
Horning D, Joyce G (2016) Proc Natl Acad Sci USA 113:9786
Houlihan G, Arangundi-Franklin S, Holliger P (2017) Acc Chem Res 50:1079
Huang Z, Szostak J (2003) RNA 9:1456
Hull W, Kricheldorf H (1980) Makoromol Chem 181:1949
Hutzinga D, Szostal J (1995) Biochemistry 34:656
Jaeger L, Wright M, Joyce G (1999) Proc Natl Acad Sci USA 96:14712
Johnston W, Umrau P, Lawrence M, Glaser M, Bartel P (2001) Science 292:1319
Joyce G (1989) Gene 82:83
Joyce G (1992) Modern Cell Biol 11:353
Joyce G (2000) Science 289:401
Joyce G (2002a) Nature 418:214
Joyce G (2002b) Science 289:401
Joyce G (2004a) Life science for the 21. century. pp 61–80.
Joyce G (2004b) Annu Rev Biochem 73:791
Joyce G, Orgel L (2006) Cold Spring Harb Monogr Arch 43:23 (RNA world. 3ed edn)
Joyce G (2007a) Science 315:1507
Joyce G (2007b) Angew Chem Int Ed 43:6420
Joyce G (2012) Science 336:307
Kay L (2000) Who wrote the book of life? A history of the genetic code. Stanford Univ. Press, Stanford CA
Kim D, Joyce G (2004) Chem Biol 11:1505
Kricheldorf H, Hull W (1980) Macromolecules 13:87
Kricheldorf H, Schilling G (1978) Makromol Chem 179:1175
Kricheldorf H, Müller D, Hull W (1985) Biopolymers 24:2113
Kricheldorf H, Müller D, Hull W (1986) Int J Biol Macromol 8:20
Koehne H, Joyce G (2003) J Mol Evol 57:292
Kuhn H, Waser J (1981) Angew Chem 93:495
Kuhns S, Joyce G (2003a) Nature 420:841
Kuhns S, Joyce G (2003b) J Mol Evol 56:711
Kumar RF, Joyce G (2003) Proc Natl Acad Sci USA 100:9738
Lam B, Joyce G (2009) Nat Biotechnol 27:288
Lam B, Joyce G (2011) J Am Chem Soc 133:23191
Lau M, Umrau P (2009) Chem Biol 16:815
Lau M, Cadieux G, Umrau P (2004) J Am Chem Soc 126:15686
Lauhon C, Szostak J (1995) J Am Chem Soc 117:1246–1257
Lawrence M, Bartel D (2003) Biochemistry 42:8748
Lawrence M, Bartel D (2005) RNA 11:1173
Lee D, Granja J, Martinez J, Severin K, Ghadiri R (1996) Nature 382:525
Lee N, Beisho Y, Weil K, Szostak J (2000) Nature Struct Biol 7:28
Lehman N, Joyce G (1993a) Nature 361:182
Lehman N, Joyce G (1993b) Curr Biol 3:723
Lincoln T, Joyce G (2009) Science 323:1229
Loakes D, Holliger P (2009) Mol Biosys 5:686

Lohse P, Szostak J (1996) Nature 381:442
Lorsch J, Szostak J (1994) Biochemistry 33:973
Luisi P (2006) The emergence of life. Cambridge Univ. Press, Cambridge
Luptak A, Szostak J (2008) In: Ribozymes and RNA catalysis. RSC Publishing, pp 123–133.
Maynard-Smith J, Szathmary E (1993) J Theor Biol 164:437
McGinnes K, Joyce G (2002) Chem Biol 9:297
McGinnes K, Wright E, Joyce G (2002) Chem Biol 9:585
McGinnes K, Joyce G (2003) Chem Biol 10:5
Merryman C, Weinstein E, Wnuk S, Bartel D (2002) Chem Biol 9:9154
Mills D, Peterson R, Spieleman S (1967) Proc Natl Acad Sci USA 58:217
Müller U, Bartel D (2003) Chem Biol 10:799
Müller U, Bartel D (2008) RNA 14:552
Mutschler A, Holliger P (2014) J Am Chem Soc 136:5193
Nielsen P (1995) Annu Rev Biophys Biomol Struct 24:167
Olea C, Joyce G (2016) Methods Enzymol 550:23
Olea C, Horning D, Joyce G (2012) J Am Chem Soc 134:8050
Olea C, Weidmann J, Dawson P, Joyce G (2015) Chem Biol 22:1437
Oparin A (1924) Proiskhozhdenie Zhizny. Izd Moskovskii Rabochii, Moscow (Reprint in transl.: Bernal 1967, pp 199–214)
Oparin A (1938) Origin of life. McMillan, Ottawa
Oparin IA (1957) The origin of life on the earth. In: Report on the international symposium 1957 (Moscow, 1957). The Publishing House of the Academy of Sciences of the UDSSR
Oparin A (1961) Life: its nature, origin and development. Oliver and Boyd, Edinburgh
Ordoukhanian P, Joyce G (2002) J Am Chem Soc 124:12499
Orgel L (1968) J Mol Biol 38:381
Pagel B, Joyce G (2010) Chem Biol 17:717
Paul N, Joyce G (2002) Proc Natl Acad Sci USA 99:12733
Paul N, Joyce G (2004) Current Opin. Chem Biol 8:634
Paul N, Springsteen G, Joyce G (2006) Chem Biol 13:329
Petrie K, Joyce G (2014) J Mol Evol 79:75
Pinheiro V, Holliger P (2012) Curr Opin Chem Biol 16:245
Pinheiro V, Holliger P (2014) Trends Biotechnol 32:321
Pinheiro V, Taylor A, Corens C, Abranov M, Renders M, Zhang J, Chapur J, Wengel J, Peak-Chew S, McLaughlin S, Herdewijn P, Holliger P (2012) Science 336:341
Pinheiro V, Ong J, Holliger P (2013a) Protein Eng Handb 3:279–302
Pinheiro V, Loakes D, Holliger P (2013b) BioEssay 35:113
Plummer KA, Carothers JM, Yoshimura M, Szostak JW, Verdine GL, (20056) Nucl. Acid Res. 33:5602–5610
Powner M, Sutherland J, Szostak J (2010) Synlett 14:1956
Reader J, Joyce G (2002) Nature 420:292
Ribas de Pouplana L (2004) The genetic code and the origin of life. Kluwer Academic & Plenum, New York
Robertson D, Joyce G (1990) Nature 344:467
Rogers J, Joyce G (1999) Nature 401:641
Rogers J, Joyce G (2001) RNA 7:395
Robertson M, Joyce G (2012) Chem Biol 21:238
Robertson M, Joyce G (2014) Cell 21:238
Sabeti P, Umrau P, Bartel D (1997) Chem Biol 4:767
Saleh-Ashitani K, Szostal J (1999) Nucl Acid Symp Ser 41:135 (Symposium on RNA biology III, RNA, tool & target)
Saleh-Ashitani K, Szostak J (2001) Nature 414:82
Santoro S, Joyce G (1997) Proc Natl Acad Sci USA 94:4262
Santoro S, Joyce G (1998) Biochemistry 37:13330
Santoro SD, Joyce G, Sahtrhival K, Gramatikova S, Barbas C (2000) J Am Chem Soc 120:2433
Sassanfar M, Szostak J (1993) Nature 364:550

Sazani P, Larralde R, Szostak J (2004) J Am Chem Soc 126:8370
Schultes E, Spasic A, Moharti U, Bartel D (2005) Nature Struct. Mol Biol 12:1130
Sczepanski J, Joyce G (2012) Chem Biol 19:1324
Sczepanski J, Joyce G (2013) J Am Chem Soc 133:13290
Sczepanski J, Joyce G (2014) Nature 515:440
Shapiro R (1987) ORIGINS. In: Bantzham (ed). Summit Books, a division of Simon & Schuster, New York
Sheppard T, Ordoulchanian P, Joace G (2000) Proc Natl Acad Sci 97:2802
Stechner D, Grant R, Bagby S, Koldobskaya Y, Piccirill J, Bartel D (2009) Science 326:1271
Stechner DM, Bartel DP (2011) Nature Struct & Mol. Biol. 18:1036–1042
Suga H, Szostak J (1998) J Am Chem Soc 120:1151
Szathmary E (1993) Proc Natl Acad Sci USA 90:9916
Szathmary E (1999) Perspective 15:223
Szathmary E, Maynard-Smith J (1997) J Theor Biol 187:555
Szostak J (1997) Chem Rev 97:347
Szostak J (2011) Philos Trans R Soc 366:2894
Szostak J (2017) Angew Chem Int Ed 56:11037
Taylor A, Pinheiro V, Smola M, Morgunov A, Peak-Chew S, Corens C, Weeks K, Herdewijn P, Holliger P (2015) Nature 518:427
Trevino S, Thang N, Elenko M, Luptak A, Szostak J (2011) Proc Natl Acad Sci USA 108:13492
Tsang J, Joyce G (1994) Biochemistry 33:5966
Tsang J, Joyce G (1996) J Mol Biol 262:31
Tuschl T, Sharp P, Bartel D (1998) EMBO J 17:2637
Umrau P, Bartel D (1998) Nature 395:260
Umrau P, Bartel D (2003) Proc Natl Acad Sci USA 100:15393
Vaas R (1994) Der genetische Code: Evolution and selbst6organisierte OptimierungAbweichungen und geziehlte Veränderungen, Wisseenschaftliche Verlagsgesellschaft, Stuttgart
Vaish N, Fraley A, Szostak J, McLaughlin L (2000) Nucl Acids Res 28(17):3316–3322
Vaish N, Larade R, Fraley A, Szostak J, McLaughlin L (2003) Biochemistry 42:8842
Voytek S, Joyce G (2007) Proc Natl Acad Sci USA 104:15288
Wang L, Schutz P (2005) Die Erweiterung des genetischen Codes. Angew Chem 117:34
Wang Q et al (2010) In: The chemical biology of nucleic acids. Wiley, p 355
Wilson C, Szostak J (1992) Curr Opin Struct Biol 2:749
Wilson C, Szostak J (1995) Nature 374:777
Wills C, Bada J (2000) The spark of life, Chapter 5. Oxford Univ. Press, Oxford
Wilson D, Szostak J (1999) Annu Rev Biochem 68:611
Witzany G (2016) Biosystems 140:49
Wochner A, Attwater J, Coulson A, Holliger P (2011) Science 332:209
Woese C (1967) The genetic code: the molecular basis for genetic expressions. Harper & Row, New York
Woriskay S, Usman N, Szostak J (2011) Proc Natl Acad Sci. USA 332:209.
Zaher H, Umrau P (2006) J Am Chem Soc 128:13894
Zaher H, Watkins R, Umrau P (2006) RNA 12:1949
Zubay G (2000) Origin of life on the Earth and in the cosmos. Academic Press, San Diego

9 The Message from Space

Space neither was, nor is pregnant with life. On the contrary, the vast majority of its molecules are highly unfavorable for a chemical evolution.

H. R. Kricheldorf

9.1 The Chemistry of Interstellar Clouds and Micrometeorites

The question, where does life on Earth come from, was answered by T. and R. Chamberlin as early as in 1908 with the speculation that reactive organic compounds delivered by meteorites and other "ambassadors" of the space were delivered to the early Earth and stimulated a chemical evolution (Chamberlin and Chamberlin 1908). In the year 1961, the American chemist J. Oro published a detailed description and discussion of the influence, extraterrestrial chemicals and forces might have had on the chemical processes on Earth (Oro 1961). According to his view, comets were the predominant vehicles that brought biochemicals to the early Earth and initiated a chemical evolution, which resulted in formation of the first cells. It was the same chemist who succeeded with the first model synthesis of a nucleobase (adenine) under quasi prebiotic conditions (s. Sect. 5.2). In the following decades, the hypothesis that the CE on Earth was initiated by "space chemicals" found more and more followers. Any kind of particles raining down from space was suspected to be effective carriers. In addition to comets, meteorites, micrometeorites, and interstellar dust particles were taken into account (Chyba and Sagan 1992; Ehrenfreund 1997; Jesenberger 1999; Pierazzo and Chyba 1999; Glavin et al. 2011; Burton et al. 2012).

In 1989, E. Anders published a calculation of the mass of organic compounds comets, meteorites, micrometeorites, and interstellar dust particles might have delivered to the early Earth (Anders 1989). According to his calculation, particles in

H. R. Kricheldorf, *Life by Chemical Evolution?*
https://doi.org/10.1007/978-3-662-70752-4_9

the mass range of 10^{-17} to 10^{-6} g should have contributed the vast majority of organic materials. He calculated a total amount of 40×10^3–50×10^3 tons p.a. Other authors objected that heating up to 1000 °C, a temperature which can be reached upon contact of the micrometeorites with the atmosphere will destroy all the organic materials. However, the fragments of larger molecules will not necessarily vanish, and thus, it cannot be denied that the micro- and macrometeorites may have made a significant contribution to the organic matter on the Earth. However, it must be considered that only a few percent of the (micro-) meteorites contain organic materials. Therefore, it is impossible to perform reliable calculations to what extent organic molecules formed in space have contributed and still contribute to the total amount of organic matter on the Earth (Love and Brownlee 1993; Taylor et al. 1998; Zolensky et al. 2006).

Numerous scientists also assumed that the number of asteroids, meteorites, and micrometeorites that hit the Earth was much higher in the beginning of the solar system, during the infancy of the Earth, so to say. This period was called the "heavy bombardment period." Furthermore, it was assumed that the surface of the Earth was not protected against UV light due to the absence of oxygen. However, severe eruptions of volcanoes have modified the atmosphere at least locally and temporarily and protected against UV light. Hence, a permanent circulation of organic molecules must have taken place between UV protected areas, where they were produced, and UV-exposed areas where they were destroyed. For all these reasons, it is impossible to estimate to what extent organic molecules reaching the Earth from extraterrestrial sources might have influenced beginning and course of a chemical evolution.

In this section and in the following one (Sect. 9.2), the question should be discussed, if organic compounds delivered by meteorites and micrometeorites to the early Earth had the quality to initiate a chemical evolution and if their quantity was high enough. A clear answer is not only important for the hypothesis of a CE on Earth, it is also relevant for the speculation, if a chemical evolution might have occurred on other planets or moons of the solar system and on exoplanets of other stars. In contrast to other books and reviews, the question as to what extent the chemicals that came in from space might have favored formation of proteins and RNAs is in the focus of this discussion.

In the course of the past 80 years, astrophysicists and astrobiologists have found out that the space between stars is not empty. It contains gases and dust particles. The concentration of all these molecules and dust particles is lower than that in a normal laboratory vacuum, but due to the huge distances, the amount of gases and condensed matter amounts to many billions of tons in our galaxy. Since new stars are born and old stars die, the concentration of organic compounds is not uniform throughout the volume of our galaxy and it varies over long periods of time. Furthermore, the entire population of organic molecules is based on an equilibration process, the efficiency of which varies from location to location. This equilibration is a consequence of the UV radiation, which can destroy any kind of chemical bonds depending on its wavelength. Molecules made up of single bonds are more frequently destroyed than molecules containing double or triple bonds. Therefore, molecules containing carbonyl, nitrile, or acetylene groups are relatively abundant among the space chemicals. In areas exposed to UV light, the ratio of destruction versus synthesis will be higher than in areas such as clouds of dust particles or in interstellar ice particles, where the molecules are more or less protected. Depending on the distance from the UV source and depending on the extent of protection by surrounding molecules (e.g., water ice), the life span of organic compounds has been estimated to vary between several months and millions of years.

Prior to detailed comments on the composition of micrometeorites and interstellar dust particles, a precise definition of these carriers from space should be given. Micrometeorites are solid particles of extraterrestrial origin having diameters in the range of 50 nm up to 2 mm. The micrometeorites that hit the surface of the Earth represent presumably <10% of the micrometeorites that touch the upper atmosphere. Due to their high speed (at least 10 km/s), they are heated up by collision with gas molecules of the atmosphere. Depending on the maximum temperature they have reached, three classes of micrometeorites were defined: unmolten particles, partially molten species, and completely molten, round, and glassy micrometeorites. Their composition is quite analogous to that of the larger meteorites. The vast majority (>90%) is bare of organic compounds and consists either of metals (mainly iron) or it contains silicates in combination with variable amounts of iron oxide. Only this small fraction of carbonaceous micrometeorites (CMMs) is of interest for the hypothesis of a chemical evolution on Earth or on other planets. A major problem is the identification of their organic components, because it is difficult to collect gram quantities without contamination by organic materials of terrestrial origin (Taylor et al. 1998).

In 2004, G. Matrajt et al. reported about amino acid analyses of micrometeorites collected in an area close to the south pole (Matrajt et al. 2004). For three reasons, they concentrated their research activities on the identification of α-amino isobutyric acid (αAibu). First, because this αAA is not a component of terrestrial peptides and proteins, and thus, its abundance on Earth is extremely low. Hence, contamination of the micrometeorites with αAibu of terrestrial origin is unlikely. Second, αAibu is chemically and thermally more stable than all prAAs and thus, its chance to survive the voyage across the atmosphere is particularly high. Third, the reaction pathway of its synthesis should be, in principle, identical with that of most prAAs, so that its detection suggests that prAAs were also formed at the location of its synthesis. αAibu was indeed detected, but only in 15% of the micrometeorites, so that an unambiguous interpretation of the results was not possible.

Finally, it should be mentioned that a small fraction of the micrometeorites contains s. c. presolar grains. This term means that these particles were formed prior to the emergence of the solar system. Hence, these presolar grains carry information about the matter formed before the sun appeared on the scene, because this matter originates from early stars that exploded via novae or supernovae. The vast majority of all micrometeorites originate, however, from components of the solar system such as comets, meteorites, and micrometeorites.

Interstellar dust particles are smaller than micrometeorites and have typically diameters <50 nm. Their composition varies over a broad range. The majority consists of silicates. But a small fraction may be aggregates of organic materials, such as oligomers, polymers, and polycyclic hydrocarbons, and they are most likely relicts of interstellar ice particles. These ice particles low their existence to the extremely low temperatures (−150 to −250 °C) which were observed for numerous interstellar regions. Below −200 °C, all gases consisting of two or more atoms crystallize. In this way, amorphous aggregates with complex chemical and physical structure may be formed. It is assumed that these frozen particles possess a solid core, most likely a silicate, which plays the role of a nucleus for the crystallization of gases. The exterior mainly consists of water and methanol mixed with smaller fractions of ammonia, formaldehyde, and carbon monoxide.

All information about gaseous or frozen molecules in interstellar clouds were retrieved by spectroscopic methods mainly based on radio waves, microwaves, or infrared radiation. In the case of interstellar ice particles, a certain fraction of the IR radiation can penetrate the amorphous outer sphere, and the absorption of certain frequencies carries information on the structure of the absorbing molecules. Electron transitions between different orbitals, rotational motions, and vibrational motions can tell the chemist what kind of chemical structure is present.

The first organic molecule, which was detected by radio-astronomy in interstellar clouds, was, methylidene (CH), a finding which was first reported in 1937. In the following years, the number of spectroscopically identified organic molecules kept steadily growing. Table 9.1 lists almost all neutral molecules detected by astronomy before 2015. The most important result with regard to chemical evolution is the finding that the diversity of molecules dramatically decreases with the number of atoms which they consist of, despite the fact that in theory the number of isomers exponentially increases with the number of atoms. Even molecules consisting of more than ten atoms are quite rare. This is a trend which is extremely unfavorable for a chemical evolution. Frankly speaking, this trend is deadly for the hypothesis of a chemical evolution initiated and sustained by space chemicals, because even a simple prAA such as phenylalanine contains 23 atoms and a nucleoside consists of 30 atoms.

Furthermore, it should be considered that a large number of molecules listed in Table 9.1 are radicals. Yet, radicals such as OH and CH_3 are highly reactive and can react with any kind of single bonds. The consequence is formation of methyl substituted isomers, a phenomenon which is discussed below in more detail. Moreover, it should be emphasized that a considerable number of molecules are ions, above all cations (Table 9.2). In the absence of solvation, these ions are extremely reactive and can initiate various series of reactions. In summary, it may be said that the interstellar matter contains a few compounds that might have been useful for a chemical evolution, but their low concentration, the strong UV radiation, and numerous reaction pathways that involve random substitution of various substrates are properties of the space chemistry which are in contradiction to an environment favoring a chemical evolution leading to a living cell.

Table 9.1 List of neutral interstellar and circumstellar molecules (incl. radicals) ordered according to their number of atoms

2 Atoms	3 Atoms	4 Atoms	5 Atoms	6 Atoms	7 Atoms
H2	CCH_2	CH_3	CH_4	$H_2C=CH_2$	$cycC_2H_4O$ (Oxirane)
CH	C_2H	C_2H_2	$CH_2=NH$	CH_3–OH	$CH_2=CHOH$
NH	HCNHNC	CH_2O	NH_2–CN	CH_3–CN	CH_3CHO
OH	$HCONH_2$	CH_2CN	$CH_2=CO$	CH_3–NC	HCC–CH_3
SH	HNO	$cycC_3H$	HCO–OH	CHO–NH_2	CH_3–NH_2
HF	C_3	$linC_3H$	HCC–CN	HCC–CHO	CH_3CH–CN
HCl	C_2O	CHCN	HCC–NC	HCCCCN	HCCCC–CN
C2	CO_2	CHCO	$cycC_3H_2$	C_5H	C_6H
CN	C_2S	HSCN	$linC_3H_2$	C_5N	
CO	SO_2	CCCN	CCCONH	$CCCCH_3$	
CSi	N_2O	C_3O	C_4H	$H_2C=C=NH$	
CP	NaCH	H_2O_2	C_4Si	$cycH_2C_3O$	
CS	MgCN		C_5	(Cyclopropenone)	
N2	SiCN		HCO–CN		
NO	AlNC				
NS	SiNC				
SiN					
PN					
SO					
O2					
SiO					
SiS					
PO					
LiH					
NaCl					
KCl					
AlF					
AlCl					

8 Atoms	9 Atoms	10 Atoms	11 Atoms	12 Atoms	13 Atoms
CH_3CO_2H	CH_3CH_2CN	CH_3COCH_3	$HC(C)_7CN$	C_6H_6	$HC(C)_9CN$
$HOCH_2CHO$	CH_3CH_2OH	$(HOCH_2)_2$	$CH_3(C)_5CH$	$CH_3CH_2CH_3CN$	
HCO-OCH_3	$CH_3(C)_3CH$	CH_3CH_2CHO			
$H_2C=CH$–CHO	$CH_3CH_2=CH_2$				
$H_2C=C=CHCN$	CH_3CO–NH_2				
NH_2–CH_2CN	C_8H,				
$HC(C)_4CH$	Ch_3OCH_3				
C_7H					

Table 9.2 Ions observed in interstellar or circumstellar clouds ordered according to their number of atoms

2 Atoms	3 Atoms	4 Atoms	5 Atoms	6 Atoms	7 Atoms	9 Atoms	>9 Atoms
CF+	H2+	HC3+	NH4+	HC3NH+	C6H–	C8H–	C60+
CH+	H2Cl+	C3N+	H2COH+				
CN+	H2O+	H3O+	HC4–				
CN-	HCO+	H2CN+	NC–CNH+				
CO+	HCS+	HCNH+					
HCl+	HN2+	HOCO+					
HO+	HOC+						

9.2 Classification of Meteorites

The analysis of extracts from meteorites is besides the radio-astronomy, the second important research strategy to retrieve information about the chemistry of the universe. Meteorites are solid matter, fragments of comets or asteroids, that survived at least partially the contact with the atmosphere and the clash on the surface of the Earth. When a meteorite enters the atmosphere, the friction with gas molecules generates heat and the evaporation of various fragments generates the fireball which allows for their observation as "shooting stars". The term meteorite is defined for solid objects having a diameter >2 mm. The largest meteorite detected so far is the s. c. Horba in Namibia. The term meteorite is not confined to cosmic objects hitting the Earth, it is also used for analogous objects that hit the Moon or Mars. Concerning the detection of meteorites, two scenarios need to be distinguished: "meteorite fall" and "meteorite find." In the former case, the fall is observed and the meteorite fragments are collected immediately after the fall. Finding of a meteorite means that its fall was not observed may have happened 10, 100, or 1000 years after its collision with the surface of the Earth. In the year 2010, 1100 "falls" were noted and 38,000 "finds."

All meteorites are classified in three categories according to their chemical composition (Krot et al. 2007):

I. Stone mete ites, which mainly consist of silicates, but usually contain a small fraction of oxidized iron.
II. Iron meteorites which mainly consist of iron (the most stable atom of all elements) and a minor fraction of nickel.
III. Stone-iron meteorites, which are composed of variable amounts of minerals and iron.

Approximately 94% of all meteorites are stone meteorites (Category I), and only 6% contain iron. About 85% of all meteorites are called chondrites, a term, which indicates that they contain large fractions of globular particles, the s. c. chondrules. All these chondrites have an age of 4.55 billions of years and they are assumed to have their origin in the asteroid belt which surrounds the Earth. A fraction of only 5% of all meteorites (typically the stone meteorites) contain a few percent of carbon compounds, and this group of meteorites is called "carbonaceous chondrites" (CCs). These CCs contain a broad variety of organic compounds (s. below) and are subdivided into eight classes. Seven classes have the symbols CX, where C stands for carbon (or carbonaceous). X is either an abbreviation of location, where they were discovered, or it originates from the most prominent meteorite of the class under consideration.

1. CB: This class is named after its most important member "Bencubbin," which was found in Australia.
2. CH: This abbreviation is an exception, because the H indicates a high metal content.

3. CI: This class has its name from the "Ivuna" meteorite found in Tansania.
4. CK: This group was named after the village Karoonda in Australia.
5. CM: The letter "M" was originally derived from the village Mighei in Ukraine, but stands today for Murchison, a small village in Australia, where the most intensively studied meteorite was found.
6. CO: This group was named after the city of Ornans in France, where the first meteorite of this class was detected.
7. CV: In this case, the village Vigevano in Italy is responsible for the name.
8. A few meteorites do not fit in with the seven classes mentioned above and have not much in common with each other. The most important representative of this group is the s. c. "Taggish-Lake" meteorite (Zolensky et al. 2002).

Finally, it is worth mentioning that the classification of meteorites also involves an Arabic number (1–6), which carries a petrographic message. A high number indicates a high extent of chemical modification of its content by water. It is the mainstream assumption that these CCs are fragments of carbon and water containing asteroids which belong to outer belt of the asteroid sphere (Morbidelli et al. 2012). Three groups, CM, CR, CI, contain organic compounds and a relatively large amount of water (3–22%). The presence of volatile molecules indicates that the meteorites of these groups were not exposed to high temperatures for a longer time. It is assumed that their composition corresponds roughly to the material which formed the basis of the solar system. In contrast to the afore-mentioned classes of meteorites, the CK, CO, and CV meteorites are almost bar of organic compounds and water. They justify the assumption that they were exposed to high temperatures even before they had come in contact with the atmosphere. Those meteorites that contained a higher-than-average amount of organic matter were intensively analyzed. Their names and classes were summarized in Table 9.3.

Table 9.3 Carbon-containing chondrites

Name	Class	Name	Class
Alan Hill	CM ½	Lonewolf Nunataks	CM 2
Allende	CV 3	Meteorite Hills	CM 1
Almahata Sitta	Ureilit	Mighei	CM 2
Bills	CM 2	Miller range	CO 3
Cold Bokkeveld	CM 2	Murchison	CM 2
Colony	CO 3	Murray	CM 2
Dominion Range	CO 3	Nogoya	CM 2
Elephant Moraine	CV 3	Orgueil	CI 1
Essebi	CM 2	Ornans	CO 3
Graves Nunataks	CR 1	Queens Alexander range	CR ½
Grosvenor Mountain	CR 1	Renazzo	CR 2
Ivuna	CI 1	Shisher	CR 2
Larkman Nunataks	CV 3	Scott Glacier	CM 1
Lance	–	Tagish Lake	–
Lewis Cliffs	CM 2	Yamato	CM 2

9.3 The Diversity of Organic Molecules in Meteorites

Fragments of all the meteorites listed in Table 9.3 were analyzed in the course of the past 50 years, but at least 80% of the information derived from these analyses are based on the "Murchison." In principle, chemical analyses of meteorites have a long tradition, because the first report of such a study was published as early as in 1834 by the Swedish chemist J. J. Berzelius (1779–1848) (Berzelius 1834). The recent analyses of meteorites concentrate mainly on the question, as to what extent their organic compounds might have contributed to the emergence of life on Earth and on other planets. Prior to a more detailed discussion of amino acids, saccharides, and nucleobases, the diversity of all organic compounds should be presented and commented. In order to avoid a lengthy listing of all pertinent original publications, a small list of representative review articles should be cited (Cronin et al. 1988; Cronin and Chang 1993; Cronin 1998; Botta and Bada 2002; Pizzarello et al. 2006; Schmitt-Koplin et al. 2010).

In the year 2006, Pizzarello et al. published a review of all the compounds extracted from the "Murchison" and from the Tagish Lake meteorites (Pizzarello et al. 2006). The results obtained from the "Murchison" are compiled in Table 9.4. A comparison with data extracted from the Tagish Lake meteorite yielded the following conclusions.

(a) Both meteorites differ largely with respect to their content of organic compounds, so that both meteorites represent two extremes of the CCs. The ratios of different classes of organic compounds relative to each other also differ largely.
(b) The Tagish Lake meteorite did not contain any amino acids.
(c) Nucleobases and nucleosides were not mentioned, but they are components of the fraction of heterocycles, as will be discussed in Sect. 9.5.
(d) The most abundant class of organic compounds were the carboxylic acids (Huang et al. 2005). Their quantity outstrips that of prAAs at least by a factor of 6, when dicarboxylic acids, hydroxycarboxylic acids, and carboxylic acids of polyols are included in this comparison. This ratio is of outmost importance for the probability that polypeptides and proteins were formed under prebiotic conditions. Carboxylic acids of any kind can terminate all kinds of polypeptide

Table 9.4 Soluble organic compound extracted from the Murchison meteorite (Pizzarello and Cronin 2000)

Group	Conc. (ppm)	Group	Conc. (ppm)
Aliphatic hydrocarbons	15	Dicarboxamides	>50
Aromatic hydrocarbons	15–28	*N*-Heterocycles	7
Carboxylic acids	>100	Amines	13
Amino acids	60	Amides	–
Imino acids	–	Polyols	30
Hydroxycarboxylic acids	15	Phosphonic acids	2
Dicarboxylic acids	>30	Sulfonic acids	67
Carboxylic acids of pyridine	>7		

syntheses, and as outlined in Sect. 3.1 such a huge excess of carboxylic acids prevents even the synthesis of higher oligopeptides.

(e) In the extracts of the "Murchison," aliphatic amines were detected (Jungclaus et al. 1976a). Their weight fractions amounts to approx. 20% of the amino acids and exceeds the weight fraction of heterocycles by 100%. Their quantity doubles when their determination is based on acidic hydrolysis of the extract. This finding suggests that most amines were originally bound in the form of carboxylic acid amides. The following primary amines were identified: methyl amine, ethyl amine, 1- and 2-propyl amine, 1-n-butylamine, sec. Butyl amine, and tert.-butyl amine.

At this point, the afore-mentioned term "polyols" should be defined. In the year 2001, Cooper et al. published a report entitled "Carbonaceous meteorites as a source of sugar related organic compounds for the early Earth" (Cooper et al. 2001). Those authors analyzed aqueous extracts of the Murray meteorite and of the "Murchsion" and found in addition to glycolaldehyde the polyols listed in Table 9.5. Along with glycolaldehyde these polyols prove that somewhere in space a s. c. "formose (or Butlerow) reaction" must have taken place. The identification of glycolaldehyde in the interstellar clouds supports this conclusion. However, this, at first glance, positive result carries a pretty negative consequence for the hypothesis of a chemical evolution on Earth. D-ribose was nowhere detected. Since D-ribose is known as a relatively unstable saccharide, this finding was not much surprising. Nonetheless, this negative result is unfavorable for most theories about the course of a chemical evolution, and it is, above all a disaster for the reductionistic-deterministic concept. Even, when a trace of ribose will be found on another meteorite, this conclusion will be correct.

Another remarkable result is the absence of alcohols in the list of Table 9.4. The author has only found one relatively old publication dealing with identification of alcohols in extracts of the "Murchison" (Jungclaus et al. 1976b). The detection of methanol, ethanol, and isomeric propanols or butanols in the aqueous extracts was predictable, since methanol and ethanol had been identified spectroscopic methods

Table 9.5 Polyols identified before 21 in extracts from the Murray and from the Murchison meteorites (Cooper et al. 2001)

Number of C-atoms	Sugars, sugar alcohols, sugar acids	Deoxy sugar acids	Dicarboxylic acids
3	Dihydroxyacetone, glycerol, glyceric a.	–	–
4	Erythrol, threol, erythronic a., threonic a.	2-methyl glyceric acid, 2,4-dihydroxy butyric a. 2,3-dihydroxy butyric a. 3,4-dihydroxy butyric a.	Tartaric acid

interstellar clouds. The significance of alcohols for the hypothesis of CE is discussed at the end of this chapter and in Chap. 11. A more recent study of diversity and weight fraction of alcohols in CCs is certainly desirable.

Finally, the comprehensive study of Schmitt-Koplin et al. (2010) on the diversity of organic species in our galaxy deserves detailed discussion. Using the most advanced chromatographic methods and mass spectrometric methods, this research group analyzed numerous different extracts of the Murchison meteorite. In addition to water, methanol, ethanol, acetonitrile, dimethyl sulfoxide, chloroform, and toluene served as solvents for the extraction. Prior to the discussion of the details, the exceptional role of the "Murchison" should be highlighted. The fall of this meteorite was documented on September 28, 1960 near Murchison, a small town in Victoria, Australia. Numerous fragments scattered over an area of 13 km^2 were collected immediately after their fall, so that contamination with chemicals or microbes was kept on the lowest possible level. All together a mass of approx. 100 kg was collected, so that in the following decades, numerous research groups had a chance to analyze the fragments. Furthermore, this meteorite proved to be CCs of the CM2 class, which was extraordinarily rich in organic compounds.

Most researchers concentrated their interest on the detection of compounds that were of particular interest for a chemical evolution and ignored substances that at first glance seemed to be of minor interest. In contrast to this one-sided behavior, it was from the beginning the purpose of the Schmitt-Kopplin group to analyze the diversity of organic compounds in all directions without considering any hypothesis or ideology. In addition to excellent technical equipment, this was the second reason why the results of this research group are of extraordinary interest and importance. Those authors summarized their results with the following words:

> Here we demonstrate that non-targeted ultra-high resolution analysis of the solvent-accessible organic fraction of "Murchison" extracted under mild conditions allows one to extend its indigenous chemical diversity to tens of thousands of different molecular compositions and likely millions of diverse structures. This molecular complexity, which provides hints on heteroatom chronological assembly, is high compared to terrestrial biochemical and biological-driven space.

Those authors also mentioned that the high diversity of isomeric compounds is characteristic for entropy-driven random substitution processes. The broad diversity resulting from random substitution processes is further enhanced by secondary chemical processes on asteroids, comets, and meteorites caused by water and/or heat. These experimental facts and their interpretation stand in diametral contrast to the visions and hypotheses that the space chemicals might have initiated a chemical evolution yielding a living cell either on Earth or on any exoplanets. Of course, the organic matter raining down from space has augmented the total amount of organic matter on Earth, but compounds that were useful or even decisive for a chemical evolution, such as optically active prAAs, nucleosides, and D-ribose were absent. In contrast, compounds such as carboxylic acids and sulfonic acids which were highly unfavorable for a chemical evolution were relatively abundant.

9.4 Identification of Amino Acids

Systematic analyses of meteorites concerning their content of prAAs began after 1960 (Kaplan et al. 1963; Hayes 1967). However, the abundance pattern was nearly identical with that of finger prints, and therefore, it was immediately assumed that the identifies prAAs resulted from contamination with terrestrial organisms such as viruses and microbes. First unambiguous results demonstrating the presence of extraterrestrial amino acids in the "Murchison" were published in 1970 by K. Kvenvolden et al. (Kvenvolden et al. 1970). That research group identified sarcosine and αAibu, which are never components of terrestrial proteins. In the following years, the existence of extraterrestrial amino acids was confirmed by several research groups (Kvenvolden et al. 1971; Oro et al. 1971; Cronin and Moore 1971; Pollock 1972; Lawless 1973; Pereira et al. 1975) (s. Tables 9.6, 9.7, 9.8, 9.9, and 9.10).

In the following decades, numerous studies of extracts from various other meteorites were published. Almost all the amino and imino acids compiled in Tables 9.5, 9.6, 9.7, 9.8, 9.9, and 9.10 were already listed in a book of U. Meierhenrich published in 2008 under the title "Amino acids and the asymmetry of life" (Meierhenrich 2008). The most surprising and most important result of all those studies is the

Table 9.6 prAAs detected in extracts of the Murchison meteorite before 2007

Name	Name
Alanine	Proline
Aspartic a.	Phenylalanine
Glutaric a.	Serine
Isoleucine	Threonine
Leucine	Tyrosine
Methionine	Valine

Table 9.7 Non-protein amino acids extracted from the Murchison meteorite

Name	Name
2-aminobutyric a.	Allo-2-amino-3-methylvaleric a.
2-amino-isobutyric a.	2-amino-2,3-dimethylvaleric a.
2-amino valeric a. (Norvaline)	Allo-2-amino-2,3-dimethylvaleric a.
2-amino-iso-valeric a.	2-amino-2,4-dimethylvaleric a.
2-amino caproic a. (Norleucine)	2-amino-3,3-dimethylvaleric a.
2-amino heptanoic a.	2-amino-3,4-dimethylvaleric a.
2-amino-2-valeric a.	2-amino-4,4-dimethylvaleric a.
2-amino-3-methylcaproic a.	2-amino-2-ethylvaleric a.
2-amino-2,3-dimethylbutyric a.	2-amino cyclopentane carboxylic a.
2-amino-2-ethylbutyric a.	2-methyl aspartic a.
2-amino-3-methylcaproic a.	3-methyl aspartic a.
2-amino-4-methylcaproic a.	Allo-3-methyl aspartic a.
2-amino-5-methylcaproic a.	2-methyl glutamic a.
2-amino-2-ethylcaproic a.	3-methyl glutamic a.
2-amino-3-ethylcaproic a.	4-methyl glutamic a.
	2-amino adipic a.
	2-amino pimelic a.

Table 9.8 w-Amino acids extracted from the Murchison meteorite

Name	Name
3-aminopropionic a. (ß-Alanine)	3-amino-3,3-dimethylpropionic a.
3-aminobutyric a.	4-amino-2-methylbutyric a.
3-amino-isobutyric a.	4-amino-3-methylbutyric a.
4-aminobutyric a.	3-amino-2-ethylpropionic a.
3-aminovaleric a.	3-amino-2,3,3-trimethylpropionic a.
4-aminovaleric a.	4-amino-3-ethylbutyric a.
5-aminovaleric a.	4-amino-2,4-dimethylbutyric a.
4-aminocaproic a.	4-amino-3,4-dimethylbutyric a.
5-aminocaproic a.	4-amino-4,4-dimethylbutyric a.
3-amino-2,3-dimethylpropionic a.	4-amino-2,3,3-trimethylbutyric a.
3-amino-2,2-dimethylpropionic a.	3-amino adipic a.
3-amino-allo-2,3-dimethyl propionic a.	

Table 9.9 Diamino acids extracted from the Murchison meteorite

Name	Name
2,3-diaminopropionic a.	3,3-diamino-isobutyric a.
2,3-diaminobutyric a.	2,5-diamino valeric a. (ornithine)
2,4-diaminobutyric a.	4,4-diamino-isovaleric a.
2,3-diamino-isobutyric a.	

Table 9.10 Imino acids extracted from the Murchison meteorite

Name	Name
N-methylglycine (sarcosine)	*N*-methylaspartic a.
N-ethylglycine	*N*-methyl-3-aminopropionic a.
N,N-dimethylglycine	Pipecoline
N-methylalanine	

finding that in the Murchison meteorite five times more non-protein amino and imino acids were found than prAAs. This means that the "Murchsion chemistry" did by no means favor a formation of prAAs. This message is a disaster for the reductionistic-deterministic hypothesis (Chap. 2, Ia) and a further unambiguous contradiction to undirected CE based on the space chemistry.

Comparisons with other meteorites underline this important conclusion. For the Murray meteorite, an amino acid composition similar to that of the "Murchison" was found. However, Yamato meteorite contains more ß-Ala and γ-Abu than all prAAs together with exception of Gly, and even αAibu is more abundant than all other prAAs with exception of Gly and Ala (Shimoyama et al. 1979). Even greater deviations from the compositions of the Murray and Murchison meteorites were found for the CI chondrites Orgueil and Ivuna (Ehrenfreund et al. 2001). The most abundant component was ß-Ala, with a concentration which was three times higher than that of Gly, and the concentration of γ-Abu was as high as that of Gly and higher than that of all other prAAs. Glavin et al. studied in 2004 the amino acid composition of the Antarctic meteorites Alan Hills and Lewis Cliffs and compared their composition with that of the "Murchison" (Glavin et al. 2004). They found significant differences, although all three meteorites belong to the same class (CM2). For example. α-Aibu was the most abundant α-AA in the case of "Lewis Cliff," and the concentration of D,L-isovaline was approximately as high as that of Gly. The concentrations of ß-Ala, γ-Abu, and α-Aibu were as high as that of Gly. In other words, most meteorites are more favorable for an evolution of polyamides than for an evolution of polypeptides and proteins. In the year 2014, A. K. Cobb and R. E. Pudlitz published a comparison of the amino acid composition of 30 meteorites. This comparison again clearly demonstrates that the space chemistry was and is far from making the dreams of CE enthusiasts coming true.

In addition to the general dispersity and relative abundance of amino acids, two aspects were of particular interest. First, the α,ω-diamino acids listed in Table 9.9 should be mentioned. U. Meierhenrich et al. reported in 2004 about the detection of these diamino acids, and although their quantities were only in the ppb range, those authors wrote the following optimistic comment (Meierhenrich et al. 2004):

> The delivery of organic compounds by meteorites interstellar dust particles or comets to the early Earth is one mechanism thought to have triggered the appearance of life on Earth.[…] In the case of the evolution of genetic material it is widely accepted that todays "RNA-Protein World" was preceded by a prebiotic system in which RNA oligomers functioned both as genetic material and as enzyme-like catalysts. In recent years oligonucleotide systems with properties resembling RNA have been studied intensively, and interesting candidates for a predecessor to RNA were found. Among these candidates a peptide nucleic acid molecule (PNA) is an uncharged analog of a standard nucleic acid in which the sugar phosphate backbone is replaced by backbone held together by amide bonds. The backbone can either be composed by *N*-(2.ethy) glycine (aeg) or its structural analog leading to aegPNA, or by diamino acids (da) leading to daPNA structures (Scheme 2.3). The nucleobases adenine and cytosine are attached via spacers to the obtained PNA structures.

The absurdity of this "Westheimer-Nielsen concepts" of a "Poly(nucleotide-peptide) World" has already been discussed in Sect. 2.3, and thus, does not need a further comment in this place.

More interesting and more important for the probability that a CE has taken place is the question, whether and to what extent the α-AAs found in meteorites possess an excess of the L-enantiomer (ee). Since it is difficult to avoid any contamination of meteorites with prAAs of terrestrial origin, it is difficult to achieve unambiguous evidence for the extraterrestrial origin of a weak "ee." Considering that L-Ala is the most abundant of all chiral prAAs extracted from meteorites and also the most abundant chiral prAAs in all model syntheses, it is not surprising that the discussion about an "ee" of the prAAs extracted from the "Murchison" started with the characterization of Ala.

In the year 1982, M. H. Engel and S. A. Nagy reported that the prAAs Ala, Asp, Glu, Leu, and Pro extracted from a Murchison fragment of the Field Museum in Chicago possess D/L ratios in the range of 0.3–0.7. The D-faction was even lower when hydrochloric acid was used for extraction. This report was in contradiction to the earlier analyses of Kvenvolden et al. (1970, 1971) and Lawless (1973) which all had found D/L ratios around 1.0. Particular surprising was the finding that two extraterrestrial amino acids, α-Aibu and α-amino-α-methyl butyric acid were fully racemized. Bada et al. (1983) criticized the data of Engels and Nagy and interpreted their results as consequence of contamination with terrestrial microorganisms. In an immediate answer (1983), Engels and Nagy repelled this critique, and in 1997, Engels and Mako published an analysis of nitrogen atoms of extracted prAAs (Engel and Mako 1997). From an extraordinary excess of the ^{15}N isotope, they concluded that their prAAs have indeed an extraterrestrial origin. But one year later (1998), Cronin and Pizzarello reported that Ala which was extracted from the interior of a Murchison fragment was fully racemized. They also presented convincing arguments that and why the results of Engels and Nagy were correct. Nearly at the same time (Cronin and Pizzarello 1997), Cronin and Pizzarello also published that all four α-methyl-α-amino acids extracted from the "Murchsion" had an "ee" around 9%. Since these α-methyl amino acids don't exist on Earth, the studies of Cronin and Pizzarello demonstrated for the first time that amino acids of extraterrestrial origin may show a clear "ee."

In the following years, the studies of α-methyl amino acids with "ee" concentrated on isovaline (Scheme 9.1), the most abundant representative of this class of amino acids. An "ee" up to 9% was confirmed for several fragments of the "Murchison" (Pizzarello et al. 2003). In the year 2009, D. P. Glavin and J. P. Dworkin published studies of the meteorites Murchison, Orgueil, Lewis Cliffs, Lone Wolf, Elephant Moraine, and Queen Alexander range (Glavin and Dworkin 2009). Using new analytical methods, those authors found "ee's" up to 18% for isovaline which was extracted from the Murchison and from the Orgueil meteorite. Surprisingly the isovaline extracted from the meteorites Elephant Moraine and Queens Alexander range were completely racemized. Pizzarello et al. concluded that a chiral degradation of α-AAS by circularly polarized UV light (predicted before) was not the only mechanism which might have generated an "ee." It was speculated that secondary reactions under the influence of water might have contributed to a slight predominance of the L-AAs, but a precise description of such processes was not provided. Yet, those authors have clearly demonstrated that secondary alterations of the organic compounds in "Murchison" were intensive. A detailed study of chemical reactions that might have influenced the composition and stereochemistry of α-methyl amino acids on board of comets and meteorites was published by Glavin et al. in 2011 (Glavin et al. 2011). In summary, all these studies definitely evidence that the extraterrestrial α-methyl amino acids may indeed possess an "ee" around or above 9%.

All authors who reported about the "ee" of α-methyl amino acids assumed that their appearance on Earth has made a significant, not to say decisive, influence on the

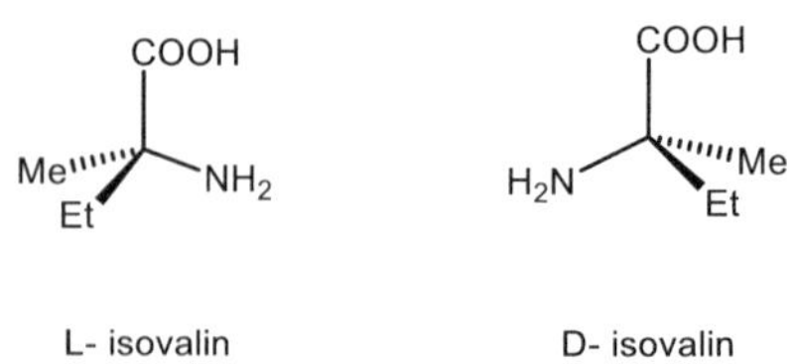

Scheme 9.1 The enantiomers of isovaline

generation of a homochiral biochemistry. Of, course, the α-methyl amino acids are not components of proteins and thus, their influence must have been an indirect one, namely in the form of catalysts. They might have played the role of ligands of metal complexes that catalyzed the formation of L-prAAs and/or D-saccharides. As discussed in more detail in the next Chapter, S. Pizzarello was able to synthesize a catalyst from isovaline with high "ee" and to use it for the preparation of saccharides having an excess of D-enantiomers. These results are certainly of great interest and an important step for a better understanding of how homochirality may be generated in lab experiments, but for several reasons, it is unlikely that isovaline having an "ee" up to 20 may have had a decisive influence on the emergence of homochirality on Earth.

1. Meteorites containing sizable amounts of amino acids represent only a small fraction among all meteorites (<1%), and their content of isovaline is extremely low.
2. The life span of isovaline and other α-methyl amino acids was short, because all energies that enabled the production of amino acids, saccharides, and nucleobases (UV light, electric discharges, etc.) also caused their destruction. The organic compounds that were produced on the early Earth did not steadily accumulate, they were part of a permanent equilibration process or cycle of construction and destruction.
3. With exception of temperatures <5 °C and pH values <7.5, prAAs racemize rapidly (s. Sect. 10.3), and therefore, any synthesis of pr AAs with high "ee" must have been endowed with high efficiency.
4. D-saccharides including D-ribose are significantly less sensitive to basic racemization than prAAs, but ribose is chemically far less stable than prAAs, so that its live span was extremely short. It fits in with this argumentation, that model syntheses allowing for an efficient synthesis of ribose were not found, and ribose has never been detected in meteorites until 2017. Even when traces of ribose will be detected in the future, this finding will not change the present argumentation.
5. The research groups of M. Powner and J. Sutherland have demonstrated that ribo-nucleosides may be prepared under alleged prebiotic conditions without intermediate formation of free D-ribose (s. Sect. 7.3).

In summary, the results of Pizzarello (2004a), Pizzarello et al. (2003) and Glavin and Dworkin (2009) are certainly important contributions to better understanding of the stereochemical aspects of the space chemistry, but it is absolutely unlikely that α-methyl amino acids have triggered the formation of a homochiral biosphere on the early Earth.

9.5 Nucleobases, Nucleosides, and Nucleotides

This section is short, because the experimental data concerning this topic are relatively meager. Analyses of meteorites with regard to their content of nucleobases, nucleosides, and nucleotides began after 1960 (Calvin 1961; Oro 1963; Kaplan et al. 1963; Hayatsu 1964; Hayatsu et al. 1968) and were continued steadily in the following five decades (Folsom et al. 1973; Vander Welden and Schwartz 1977; Hayatsu et al. 1975; Stocks and Schwartz 1979, 1981; Martins et al. 2008; Callahan et al. 2011). The identification of nucleobases was a great analytical challenge, because of their low concentration, the presence of many other compounds, and the high risk that contamination with terrestrial microbes or viruses falsified the results. This situation has been described by Callahan et al. in the introduction of their publication "Carbonaceous Meteorites Contain a Wide Range of Extraterrestrial Nucleobases" (Callahan et al. 2011) as follows:

> To date all the purines [A, G, xanthine and hypoxanthine, s. Scheme 5.2) and the one pyrimidine (uracil) reported in meteorites are quite common and could be explained by terrestrial contamination. Martins et al. performed compound specific carbon isotope measurements for uracil and xanthine in the Murchison meteorite and int4erpreted the isotope signatures as non-terrestrial. However, other co-eluting molecules (e.g., carboxylic acids known to be extraterrestrial) could have contributed to the d12C values for the nucleobases. Furthermore, there have been no observations of stochastic molecular diversity of purines and pyrimidines in meteorites which has been a criterion for establishing extraterrestrial origin for nucleobases detected in carbonaceous chondrites never been established unequivocally, nor has the detection of nucleobases been demonstrated in more than a handful of meteorites.

The first report on the detection of a nucleobase in a meteorite (i.e., cytosine) (Calvin 1961) was later identified as mistake, based on an artifact of the analytical method. (Oro 1963, Kaplan et al. 1963). Later, a report of Hayatsu et al. (1975) about the identification of unusual nucleobases bare of any biological function was also shown to be incorrect (Vander Welden and Schwartz 1977). However, in the following years, several research groups confirmed that several CM2 chondrites contained the purines adenine, guanine, xanthine, and hypoxanthine. Uracil was only detected in the "Murchison" and only in difficult to identify traces. None of those reports contains information about presence or absence of biologically non-relevant nucleobases. The studies of those authors were fully focused on identification of nucleobases present in RNA and DNA. Such a one-sided kind of research has the short-coming that it may provide a distorted picture of the diversity of nucleobases and similar heterocycles in our galaxy. In 2011, Callahan et al. started a search for non-biological nucleobases and discovered purine and 2,6-diamino purine in the extracts of "Murchison" and "Lonewolf Nunataks" (Callahan et al. 2011).

Considering the omnipresence of methyl and ethyl radicals in interstellar and circumstellar clouds and as consequence the identification of numerous methyl and ethyl-substituted amino acids and carboxylic acids, it is strange that no methyl and ethyl-substituted nucleobases were detected. Furthermore, any information on presence or absence of other N-heterocycles, such as pyrrole, imidazole, indole, and benzimidazole (s. Scheme 5.2) and their methyl or ethyl derivatives is lacking. Provided that the chemical evolution proceeded via the reaction between ribose and nucleobases, all these N-heterocycles had competed with the biological nucleobases for glycosidation by ribose. Such a competition had the consequence that formation of RNAs was almost impossible quite analogous to formation of proteins in the presence of numerous non-protein amino acids.

In 2015, B. Pearce and R. Pudlitz published a review on the relative abundance of nucleobases in 14 meteorites (Pearce and Pudlitz 2015). The purine bases A and G were found in all cases, but their weight fraction was in the range of 1–103 ppb corresponding to 1% relative to Gly. Uracil was only detected in the "Murchison," whereas cytosine and thymine were nowhere detected at that time. However, in 2022 and 2023, two communications appeared which improved the situation. At first, all canonical nucleobases including cytosine and thymine were detected in extracts of Murchison (Oba et al. 2022), and uracil was detected on the carbonaceous asteroid Ryugu (Oba et al. 2023). For two reasons, these new results do not support the theory that space chemistry has favored a chemical evolution on Earth or exoplanets. First, the detection of the canonical nucleobases was based on improved analytical methods which allowed for the detection of quantities as low as 4 ppb's (of the extracted material). Yet, such extremely low quantities cannot be taken serious as initiators or boosts of a CE on Earth. Second, as mentioned in Chap. 5, no plausible pathway was found explaining, how nucleosides were formed from nucleobases and D-ribose under prebiotic conditions. These negative aspects are further aggravated by the fact that neither nucleosides nor nucleotides were detectable. The absence of nucleosides and nucleotides may be understood as consequence of the absence of ribose. However, the research groups of M. Powner and J. Sutherland have recently demonstrated that it is possible to prepare nucleotides starting out from simple components of interstellar clouds such as glycolaldehyde, HCN, and cyanoacetylene. Yet, when nucleosides and nucleotides are absent from the space chemistry, it is definitely clear that this space chemistry did never had the tendency to support any chemical evolution neither on Earth nor elsewhere.

With regard to the cited, famous statement of Ch. DeDuve "Space was pregnant with live" the author has coined an alternative statement at the beginning of this chapter which is in agreement with the experimental facts.

9.6 Amphiphilic Membrane Forming Compounds

Cells are characterized by formation of a membrane which enables them to perform in their interior chemistry which is different from that outside, but permits a selective transport of chemicals across the membrane. Vesicles consisting of membranes that have a sufficient mechanical stability in combination with semi-permeability are the most promising candidates as precursors of cells known from living organisms. Organic molecules that possess such a combination of properties are known as amphiphilic compounds what means that they are composed of a hydrophilic and a hydrophobic group (s. Fig. 2.3). The simplest representatives of this class of compounds are carboxylic acids having a long aliphatic chain, meaning an unsubstituted alkane chain consisting of more than 10 CH_2 groups. The stability of the membrane decreases with the number of CH_2 groups and even more dramatically with the number and size of (alkyl)substituents. Cis C=C double bonds also destabilize the membrane, whereas one trans double bond does not have significant influence.

In addition to aliphatic carboxylic acids, long chain alkanols may be components of amphiphilic compounds, provided that the O-atom is attached to a hydrophilic group such as phosphoric acid or oligoethylene oxides. A couple of amphiphilic compounds which are characteristic for living organisms are compiled in Scheme 9.2.

Here, the question should be answered, whether and to what extent such membrane forming amphiphiles were extracted from meteorites. First studies in this direction were described by D. W. Deamer (Deamer 1985, 1986, Deamer and Pasiley 1989). Fragments of the "Murchison" were extracted with chloroform, followed by fractionation of the extracts. For the characterization, the following methods were applied: UV- and IR spectroscopy, optical microscopy, acid-base titration, and measurements of the surface tension. As surface active compounds, alkanols and phenols were identified, with Deamer's own words:

> Thus certain components present in the Murchison chondrite have the ability to form two kinds of boundary structures at alkaline pH range. Microstructures in the form of viscous fluid droplets are analogous to the original "coacervate" concept of Oparin is that they are complex mixtures of hydrated organic substances which form a gel, when interacting with aqueous phases [….]. The second boundary structure takes the form of actual membranes, defined here by a thin boundary capable of acting as barrier to free diffusion.

In the article of 1989 Deamer says:

> In contemporary membrane lipids the hydrophobicity is provided by hydrocarbon chains of fatty acids with chain lengths in the range of 12 to 20 or more carbons. A plausible source in the prebiotic environment is not readily apparent and it is significant that such chains are relatively abundant in carbonaceous chondrites. Smith and Kaplan (1970) demonstrated that chain lengths of 13–23 carbons in seven different CM2 chondrites and a similar distribution was observed in the "Murchison" (Lawless and Yuen 1979).

Scheme 9.2 Examples of selected natural lipids

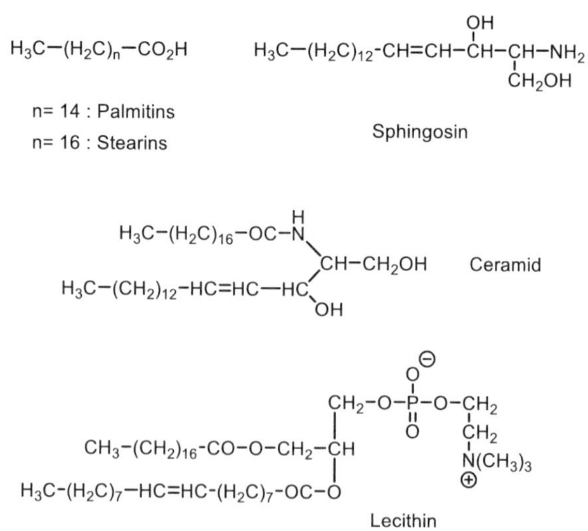

This comment sounds so, as if Deamer believed that the mentioned aliphatic carboxylic acids were all unsubstituted *n*-alkanoic acids that were delivered by comets and meteorites to the early Earth, where no reaction pathway for their synthesis existed. However, neither Deamer nor Smith and Kaplan analyzed the exact structure of those aliphatic carboxylic acids. Lawless and Yuen (1979) have analyzed and demonstrated that the carboxylic acid they had extracted were mixtures of unsubstituted and substituted species, with a greater abundance of the substituted ones. The same authors have later performed model syntheses of aliphatic carboxylic acids (1981), whereupon the Miller-type electric discharge experiments were used. The distribution of substituted and unsubstituted acids was nearly the same as that extracted from the "Murchison." Two trends were found:

1. The molar concentration of a carboxylic acid decreases with their molar mass. This trend fits in with the distribution of molar masses of various molecules in interstellar clouds as illustrated in Table 9.1.
2. The molar ratio of *n*-alkanoic acids relative to their branched analogs decreases with increasing number of carbon atoms. This is a normal consequence of random substitution with methyl or ethyl radicals.

All these results agree well with the predominance of UV-driven radical chemistry in space as demonstrated by the results of Schmitt-Koplin et al. (2010). Scheme 9.3 illustrates for a radical carboxylation of octane that for statistical reasons branched octanoic acids must predominate in the reaction products. Furthermore, it must be taken into account that radicals are more easily formed by CH_2 groups than by CH3 groups, because the resulting secondary radical is more stable than the primary radical. Therefore, no experimental evidence has been found that the linear alkanoic acids or n-alkanols needed for formation of the membranes known from

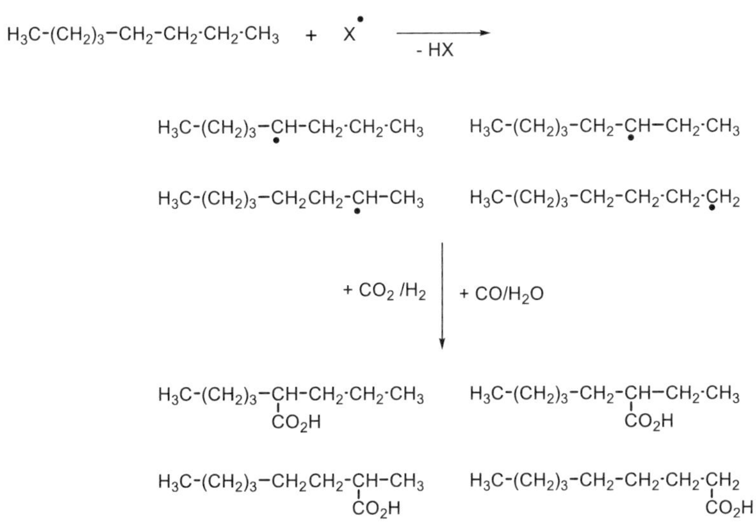

Scheme 9.3 Formation of isomeric nonanoic acid through statistical radical substitution at octane

the cells of living organisms were contributed to chemical evolution on Earth from the space chemistry. Moreover, no model syntheses were elaborated explaining how these amphiphilic compounds might have been formed on the earth itself. It is known that reaction mixtures mainly showing a predominance of *n*-alkanoic acids over branched analogs may be obtained via the Fischer-Tropsch process, which is based on the reaction of CO with H_2 at high pressure (25 bar) and high temperatures (20–300 °C) and requires special iron or cobalt catalysts to avoid still higher temperatures and pressures. Yet, such harsh conditions are a highly unlikely scenario for the early Earth due to the absence of hydrogen and high pressure. Another approach to n-alkanoic acids was speculatively formulated by Zubay (2000). That author assumed that polycondensation of formaldehyde mainly resulted in linear polysaccharides, which upon dehydration produced linear unsaturated carboxylic acids. The dehydration was in turn, somehow followed by a hydrogenation process. This hypothesis is a typical example of a speculation, where the characterization as absolutely unlikely is still euphemistic.

Finally, three reports about the detection of organic compounds in space should be cited, which are typical examples of overoptimistic interpretations of the correlation between space chemistry and chemical evolution on Earth. The astrophysicists N. F. W. Likterink et al. reported in 2017 that identification of methyl isocyanate and its importance for the CE as follows (Likterink et al. 2017):

> Complex organic compounds defined in astrochemistry as molecules that consist of six or more atoms are widely found in star-forming regions. A special category of special molecules is that of the prebiotics, molecules that can be linked via their chemical structure or reactivity to the live-bearing molecules, such as amino acids or sugars. Methyl isocyanate (CH_3NCO) also known as isocyanatomethane is a molecule that falls in that category because of its structural similarity with a peptide bond.

This comment puts the understanding of structure–reactivity relationships of an organic chemist upside down. For a chemist neither structure nor reactivity of methyl isocyanate resembles those of a peptide group. Consider methyl isocyanate was present on Earth in quantities sufficient to influence the ongoing chemical processes, then its influence was absolutely negative. First, isocyanate reacts rapidly with primary aliphatic amino groups yielding urea groups and thus, would have prevented any kind of protein syntheses. Second, its reaction with amino groups of nucleotides, even if one with low conversion, would have hindered their biological function. Third, the methylamine resulting from hydrolysis of methyl isocyanate had competed with the primary amino groups of amino acids and hindered any synthesis of proteins. Fourth, under prebiotic conditions yielding prAAs, the presence of methyl amine yields sarcosine and other N-methyl amino acids as demonstrated by the experiments of S. L. Miller and by the extracts of "Murchison." These N-methylamino acids are, in turn, highly unfavorable as components of proteins with biological function, because they destabilize any ordered secondary structure.

The second example concerns a report of P. Gasda et al. (2012) dealing with identification of borates on the planet Mars. It was speculated that the boric acid has stabilized D-ribose via formation of a cyclic ester (involving the OH groups in

2'and 3'position) and thus favored the formation of nucleosides and finally the emergence of life. At least four arguments may be forwarded against this overinterpretation. First, borates, i.e., metal salts of boric acid, don't react with ribose at moderate conditions, free boric acid or their esters are needed for esterification of ribose. Second, boric acid can form cyclic esters with any kind of 1,2-diols, when the OH groups can adopt a cis position/conformation. Ribose is not an exception. Third, ribose was neither detected in comets nor in meteorites nor in interstellar clouds. Fourth, under conditions where the cyclic borates are stable, they prevent formation of cyclic phosphates which may yield RNAs by polymerization, and they affect the polycondensation of nucleoside 5′-phosphates.

The third example is a report on Brunken et al. (2022) about the first detection of dimethyl ether in planet forming disks. The abstract of their paper begins with the sentence: "The complex organic molecules detected in star-forming regions are the precursors of the prebiotic molecules that can lead to the emergence of life." Yet, how dimethyl ether may have contributed the formation of prAAs or nucleosides remain obscure.

These examples demonstrate how researchers try to underline the importance of their discoveries in that they emphasize that the detected compounds may have had a favorable influence on the course of a chemical evolution toward a first living cell. In this way, the myth of a space chemistry favoring the emergence of life is continued and supported corresponding to De Duve's statement that "space was pregnant with life"….(DeDuve 2002, s. Hypothesis I, Sect. 1.1).

References

Anders E (1989) Nature 342:255–257

Bada JL, Cronin JR, Hu M-S, Kvenvolden KA, Lawless JG, Miller SL, Oro J, Steinberg S (1983) Nature 302:494–496

Berzelius JJ (1834) Ann Phys Chem 33:134

Botta O, Bada JL (2002) Surv Geophys 23:411–465

Brunken N, Booth A, Leemker M, Nazari P, van der Marel L, van Dishoek E (2022) Astron Astrophys 659:A29

Burton AS, Stern J, Elaila JE, Glacvon DP, Dworkin JP (2012) Chem Surv Rev 2012:54–59

Buseman H et al (2006) Science 312:727

Callahan MR, Smith KE, Cleaves II, Ruzika J, Stern JG, Glavin DP, House CH, Dworkin JP (2011) Proc Natl Acad Sci USA 106:1295–1399

Calvin M (1961) Chem Eng News 39:96–104

Chamberlin T, Chamberlin R (1908) Science 28:897–910

Chyba CF, Sagan C (1992) Nature 355:125–132

Cobb AK, Pudlitz RE (2014) Astrophys J 783:140

Cooper G, Kimmich N, Belisle W, Saumana J, Brabham K, Garell L (2001) Nature 424:879–883

Cronin JR (1998) In: Brack A (ed) The molecular origin of life: assembling pieces of the puzzle. Cambridge University Press, Cambridge

Cronin JR, Chang S (1993) In: Greenberg JM et al (eds) The chemistry of life's origin. Kluwer Academic Press, Dordrecht, pp 209–258

Cronin JR, Moore CB (1971) Science 172:1327

Cronin JR, Moore CB (1997) Science 172:1327

Cronin JR, Pizzarello S (1997) Science 275:951–995

Cronin JR, Pizzarello S, Cruikschank DP (1988) Orgabic matter in carbonaceous chondrites, planetary satellites, asteroids and comets. In: Kerridge JF, Matthews MS (eds) Meteorites and the early solar system. University of Arizona, Tucson, pp 819–857
Deamer DW (1985) Nature 265:78
Deamer DW (1986) Orig Life Evol Biosph 17:3
Deamer DW, Pasiley RM (1989) Orig Life Evol Biosph 19:21
DeDuve C (2002) Life evolving, molecules, mind and meaning. Oxford University Press, Oxford
Ehrenfreund P (1997) Science 283:1123–1124
Ehrenfreund P, Glavin DP, Cooper G, Bada JL (2001) Proc Natl Acad Sci USA 98:2138–2141
Engel MH, Mako SA (1997) Nature 389:265–268
Engel MH, Nagy B (1982) Nature 296:837
Engel MH, Nagy B (1983) Nature 301:496
Folsom CE, Lawless JG, Ro M, Ponamperuma C (1973) Geochim Cosmochim Acta 37:455–465
Gasda PJ, Haldeman EB, Wiens RC, Rapin W, Bristow TF, Bridges JC, Schwarzer SP, Clkark B, Herkenhoff K, Frydenwang SD, Lanza NL, Maurice S, Clegg S, Delapp DM, Sanford VL, Bodine MR, McIntry R (2012) Geophys Res Lett 44:8739–8748
Glavin DP, Callahan P, Dworkin JP, Elsila JG (2011) Meteorit Planet Sci 45:1948
Glavin DP, Dworkin JP (2009) Proc Natl Acad Sci USA 106:5487
Glavin DP, Dworkin JP, Aubrey A, Botta O, Doty III JH, Martins Z, Bada JL (2004) Meteorit Planet Sci 101:9181–9186
Hayatsu R (1964) Science 146:1291
Hayatsu R, Shidiev MH, Moore LP, Anders E (1975) Geochim Cosmochim Acta 39:471
Hayatsu R, Studier MH, Matsuoka S, Anders E (1968) Geochim Cosmochim Acta 38:175
Hayes JM (1967) Geochim Cosmochim Acta 31:1395
Huang Y, Wang Y, Alexander MR, Lee T, Petruck C, Fuller M, Pizzarello S (2005) Geochim CVosmochim Acta 89:1073–1084
Jesenberger EK (1999) Space Sci Rev 90:91–92
Jungclaus GA, Cronin JR, Moore CB, Yuen GU (1976a) Nature 261:121–128
Jungclaus GA, Cronin JR, Moore CB, Yuen GU (1976b) Meteoritics 11:231–237
Kaplan IR, Degens ET, Reuter JH (1963) Geochim Cosmochim Acta 27:805–834
Krot AN, Keil K, Scott ERD, Goodrich CA, Weisberg MK (2007) In: Holland HD, Turekian KK (eds) Treatise on geochemistry. Elsevier, Amsterdam, pp 83–128
Kvenvolden K, Lawless JG, Pering K, Petersen E, Flores J, Ponamperuma C, Koplin IR, Moore C (1970) Nature 228:923–926
Kvenvolden K, Lawless JG, Ponamperuma C (1971) Proc Natl Acad Sci USA 68:486–490
Lawless JG (1973) Geochim Cosmochim Acta 37:2207
Lawless JG, Yuen G (1979) Nature 282:396
Likterink NFW, Coutens A, Kofman V, Müller HSP, Garrod RT, Calcun H, Wampfler SF, Jorgensen JH, Limartz H, van Dishoek EF (2017) MNARS (Advanced Access publ) 469:2219
Love SG, Brownlee DE (1993) Science 262:550–553
Martins Z, Botta O, Fogl ML et al (2008) Earth Planet Sci Lett 270:130–135
Matrajt G, Pizzarello S, Taylor S, Brownlee D (2004) Meteorit Planet Sci 39:1840–1858
Meierhenrich U (2008) Amino acids and the asymmetry of life, Chapter 10.6. Springer, Berlin
Meierhenrich U, Munoz-Caro G, Bredehöft JH, Jesseberg E, Thiemann W (2004) Proc Natl Acad Sci USA 101:9182
Morbidelli A, Lumine J, O'Brien D, Raymond S, Walsh K (2012) AREPS 40:251
Oba Y, Takano Y, Furukawa Y, Koga T, Glavin D, Dworkin J, Naraoka H (2022) Nat Commun 13:2008
Oba Y et al (2023) Nat Commun 14:1292
Oro J (1961) Nature 190:389–390
Oro J (1963) Nature 197:756–758
Oro J, Lichtenstein S, Wikstrom S, Flory DA (1971) Nature 230:105
Pearce BKD, Pudlitz R (2015) Astrophys J 807(85):1–10
Pereira WE, Summons R, Rindfleich TC, Duffield AH, Zeizman B, Lawless JG (1975) Geochim Cosmochim Acta 30:163

Peters Z, Botta O, Charnley SB, Rutterkamp R, Ehrenfreund P (2003) Astrophys J 593:L125
Pierazzo E, Chyba CF (1999) Meteorit Planet Sci 34:900
Pizzarello S (2006) Acc Chem Res 39:231
Pizzarello S, Cooper GW, Flynn GJ (2006) In: Lauretta DS, McSween HJ (eds) Meteorites and the early solar system II. University of Arizona Press, Tucson
Pizzarello S, Cronin JR (2000) Geochim Cosmochim Acta 64:329–338
Pizzarello S, Weber AL (2004a) Science 303:1151
Pizzarello S, Zolansky M, Turk KA (2003) Geophys Cosmochim Adta 67:1589–1595
Pollock GE (1972) Anal Chem 44:2368
Remusat F, Darenne S, Roberts S, Krucher H (2005) Geochim Geophys Acta 69:3919
Schmitt-Koplin P, Gabelica Z, Gaugeon RD, Fekete A, Kanawati B, Harir M, Gebefuegi E, Eckel G, Hertkorn N (2010) Proc Natl Acad Sci USA 107:2763–2768
Sephton MM (2002) Nat Prod Rep 19:292–311
Shimoyama Y, Ponamperuma C, Yannik K (1979) Nature 282:394–396
Smith JW, Kaplan JR (1970) Science 167:1367
Stocks PG, Schwartz AW (1979) Nature 282:709–710
Stocks PG, Schwartz AW (1981) Geochim Geophys Acta 45:563–569
Taylor S, Lever J, Harvey RP (1998) Nature 392:899–903
Taylor S, Schmitz JH (2001) In: Peuker-Ehrenbrink BN, Schmitz B (eds) Accretion in extraterrestrial matter throughout earth's history. Kluwer Academic Press, New York
Vander Welden W, Schwartz AW (1977) Geochim Cosmochim Acta 41:961–968
Zolensky M, Bland M, Browa P, Halliday I (2006) In: Lauretta DS, McSween H (eds) Meteorites and the early solar system. University of Arizona press, Tucson
Zolensky M, Nakamura K, Gounelle M, Mikouchi T, Kasamura T (2002) Meteorit Planet Sci 37:737–761
Zubay G (2000) Origins of Life on the earth and in the cosmos. 2nd edn. Academic Press, San Diego

The Problem of Homochirality

10

The great tragedy of Science: the slaying of a beautiful theory by an ugly fact.

T. H. Huxley

10.1 Origins of an Enantiomer Excess

The biopolymers of all living organisms that were analyzed in the course of the past 150 years consist of homochiral monomer sequences, what means that all monomer units belong to the same class of enantiomers. Peptides and proteins of most multicellular organisms consist of L-prAAs, and only several fungi and microorganisms synthesize small amounts of peptides containing one or (rarely) two D-amino acids. The poisonous cyclic peptide phalloidin produced by the mushroom *amanita phalloides is the most widely* known example of a peptide containing a D-amino acid (D-Thr). All polysaccharides and the ribose and deoxy-ribose units in RNA and DNA exclusively consist of D-sugars. Poly(ß-hydroxybutyrate), a polyester produced as food reserve by numerous bacteria also consist of D-units. Natural rubber, which does not contain chiral building blocks, is made up by a perfect sequence of one geometrical isomer, the cis-isoprene unit, whereas other plants produce perfect poly(trans-isoprene).

Why are sequences of stereochemically identical repeat units so important? A first simple answer is to say that the stereochemical uniformity is an optimization of efficiency and functionality. The aspect of efficiency can, for example, easily be demonstrated in the case of a polypeptide having a length of 50 prAA units and two different end groups. All possible stereoisomers that can be formed from an equimolar D/L ratio amount to 250 stereocopolypeptides, whereas in the absence of D-amino acids, only one polypeptide can be formed. In this way, a tremendous simplification for the pool of biochemically useful polypeptides and proteins will result. A further simplification is the consequence that only one sort of enzymes is

H. R. Kricheldorf, *Life by Chemical Evolution?*
https://doi.org/10.1007/978-3-662-70752-4_10

needed to synthesize or hydrolyze these stereochemically uniform biopolymers. Furthermore, homochiral polymers can undergo more specific and more stable interactions with other chiral or homochiral molecules regardless if the reaction partners are monomers, oligomers, or polymers. For this aspect, it does not matter if the interaction is based on partial valencies, metal complexes, ionic bonds, or covalent bonds.

Several authors apply the terminology typical for the stereochemistry of polyolefins and vinyl polymers also to biopolymers. In this terminology, isotactic means homochiral sequence, syndiotactic means alternating sequence of D- and L-units, and atactic means random stereosequence. However, such a mixing of different terminologies does not make much sense, because the characteristic differences between both terminologies may cause misunderstandings as demonstrated by the following points:

1. In the case of biopolymers, each monomer is chiral, whereas the common α-olefins and vinyl monomers are prochiral.
2. Biopolymers possess a polarity along their chains and never have two identical end groups.
3. Polyolefins and vinyl polymers can form helices in the solid state, but these helices are unstable in solution. Furthermore, enantiomeric helices (helices with opposite twist) are mirror images and have identical thermodynamic properties (enthalpies and entropies). In contrast, helical biopolymers having a left or right twist are not mirror images, and they have a diastereomeric character and possess different thermodynamic properties and different reactivities.

Furthermore, it is worth noting that the R-/S nomenclature widely used in organic and inorganic chemistry can in principle applied to biomonomers, but for the following reason the substitution of the classical D/L nomenclature does not make sense. The stereochemical rules underlaying the D/L-nomenclature are quite different from the seniority rules underlaying the R/S-nomenclature, so that R and S do not exactly parallel to D and L. For example, 19 prAAs having the L-configuration may be labeled S-amino acids, but in the case of Cys and cystine, the L-form requires the R-label.

An important fundamental law of chemistry says that the synthesis of a chiral compound from achiral starting materials in an optically inactive environment necessarily yields racemates. His law of nature was confirmed in countless experiments and the syntheses of prAAs by Miller and coworkers (Chap. 4) are a characteristic example. For a long time, this law seemed to be in perfect agreement with various physical laws. Chiral parity seemed to be an unavoidable consequence of all chemical and physical experiments under isotropic conditions. Therefore, it was a big surprise, when in the late 1960s, Glashaw, Salan, and Weinberg published a unified theory of electromagnetism and weak interaction, the s. c. standard model, which predicted the possibility of parity violations (Latal 1992).

This change of paradigm in physics had important consequence for the understanding of chirality in chemistry and in the hypothesis of chemical evolution (Janoschek 1992). Two important results should be mentioned. The quantum

Table 10.1 Energy barriers and times needed for racemization determined by spectroscopic methods for tunneling (Janoschek 1992)

Molecule	Energy barrier (Kal/mol)	Rac. Time (s)
NH3	5	2.1×10^{-11}
PH3	37	17.0×10^{7}
CH4	112	$\sim 10^{25}$

mechanics yielded achiral eigenfunctions the overlapping of which predicted left and right-handed enantiomers which might be understood as time-dependent states of a chiral molecule. Both states are separated by an energy barrier which may be relatively low as it is typical for tertiary amines having three different substituents. In this case, the thermal energy available at room temperature suffices to cause a rapid exchange between both enantiomers. When the energy barrier is high as it is true for most sp^3-hybridized carbon atoms, and racemization is extremely slow and the life time of an enantiomer may amount to thousands or even to millions of years (Table 10.1). For relatively stable enantiomers, in principle, three racemization mechanisms exist:

1. Tunneling
2. Thermal activation by heating
3. Chemical catalysis

The most surprising aspect of the standard model is the prediction that irrespective of the molecular structure both enantiomers do not possess the same energy content (parity violation energy difference, PVED). This energy difference is extremely small and amounts for prAAs to only 10^{-17} Kcal/mol (Janoschek 1992). Mason and Tranter (1985) calculated slightly higher value of 10^{-14} Kcal/mol. For prAAs, the calculations according to the standard model yield a higher stability of the L-enantiomer, but for saccharides a higher stability of the D-enantiomer.

Hence, these calculations perfectly agree with the fact that all living organisms consist of L-prAAs and D-saccharides. What means, that both the world of elementary particles and the world of biomolecules are both inherently chiral and asymmetrical. Despite this amazing analogy, it was not possible so far to explain, how the properties of the elementary particles were transformed into the homochirality of biopolymers in the course of a chemical evolution.

However, this agreement cannot be used as prove for the existence of a chemical evolution. It may also be understood as consequence of a consistent design of the entire universe when the first cells appeared by singular sudden event (or creation). According to mason and Tranter, the PVED of 10–14 Kcal/mol corresponds to an excess of 10^6 L-enantiomers among the 10^{23} amino acids of 1 mol or in other words a molar ratio of $1{:}10^{17}$. This low ratio is, in turn, meaningless for chemical experiments in laboratories and industry, and a relevance for the course of a chemical evolution has never been demonstrated. It remains the fact that in the macroscopic world, racemization is thermodynamically favored, because it represents a gain of entropy.

Tranter proposed in 1985 that one chiral quartz crystal in excess of 10^{17} equimolar amounts of enantiomeric quartz crystals might have sparked a permanent increase of one enantiomer via stereoselective or enantioselective catalytic processes (Tranter 1985). A similar line of argumentation was published in 1991 by Wächtershäuser for catalytic activities of chiral pyrite crystals (Wächtershäuser 1991). A more recent hypothesis of Jakschitz and Rohde (2012) takes into account that the PVED dramatically increases with the order number of an atom, so that the PVED of prAA copper complexes is three orders of magnitude higher than that of the neat prAA. For syntheses of oligopeptides involving Cu-complexes of prAAs, a slightly higher degree of homochiral coupling was observed than for analogous experiments in the absence of Cu ions. However, as demonstrated in Sect. 10.3, homochiral coupling is sensitive to various parameters, and chelate complexes of prAAs have the disadvantage that they favor racemization of the prAA ligand by a factor of 10^2–10^3.

To find an explanation how a world of various living organism based on homochiral biopolymers might have emerged from a prebiotic world of racemates, two different scenarios have to be considered and investigated.

I. D-amino acids and L-saccharides were absent from the beginning or have totally disappeared before biopolymers were formed.
II. Starting out from a small enantiomeric excess of a L-prAA or a D-saccharide, the homochiral biopolymers were formed in a series of enantioselective (most likely autocatalytic) and homochiral coupling-favoring reactions. An efficient cascade of such autocatalytic reactions was recently calculated by Huber and Trapp (2022), but the temperature dependence of the reaction rates was not considered and such processes have to fast and efficient below 25 °C, otherwise they cannot compete with racemization (s. Sect. 10.3).

Scenario (I) is extremely unlikely because of the high temperatures before and at the beginning of the chemical evolution. Temperatures above 50 °C and a slightly basic environment as it is typical for the oceans today cause rapid racemization of prAAs as discussed below (Sect. 10.3). Due to the permanent racemization, a singular event such as impact of an asteroid or the arrival of several carbonaceous meteorites within a few hundred years has not changed the chiral scenario on the early Earth. A strong permanent invasion of L-amino acids and D-saccharides from space in combination with low temperatures (<10 °C) might have outpaced racemization, but there is no evidence that such a scenario has existed on Earth.

For scenario (II), the processes that might have resulted in an enantiomer excess should here be discussed under four aspects.

(a) Processes that have favored the ab initio formation of one sort of enantiomers.
(b) Processes that might have favored the destruction of one sort of enantiomers.
(c) Processes that caused an enantioselective enrichment.
(d) Condensation and polymerization processes that favored homochiral coupling and in combination with selective degradation of heterochiral sequences indirectly favored enrichment of one sort of enantiomers. This aspect will be discussed in more detail in the next section.

Ad (a): After World war II research activities dealing with syntheses, chromatographic separation and identification of optically active chiral molecules rapidly developed in a new and diversified working field (Faber and Griengel 1992; Winterfeld 1992; Jacobsen et al. 1999; Berkessel and Groyer 2005; Glorius and Glas 2006; Faber 2011; Bauer 2012). Stereoselective syntheses of medicaments were the main driving force for this evolution, because the biochemical function, and thus, the usefulness as medicament are usually to only one enantiomer. The mirror image is either inactive or poisonous. The Nobel Prizes of 2001 awarded to W. S. Knowles, R. Noyorri, and K. B. Sharpless underline the importance of this working field. For the transformation or enantio-selective synthesis of one enantiomer, optically active catalysts were used which in almost all cases were natural products.

Among the numerous stereo-selective synthesis that were elaborated during the past 60 years, a few syntheses of optically active amino acids can be found. Yet, none of these syntheses shed light on the question, how under prebiotic conditions, optically active prAAs might have been formed from achiral or prochiral starting materials. The only kind of chiral matter that was thermostable and available in large amounts on the early Earth were quartz crystals. A sequence of SiO_4 tetrahedra can form a left- or right-handed helical array, so that even the surface of quartz crystals may have a chiral character. Metal ions or metal complexes adsorbed on the surface of such quartz crystals may, therefore, have the properties of chiral catalysts which catalyze enantioselective reactions.

First experiments in this direction were first described by Klabunowski and coworkers (Terent'ev and Klabunowsku 1953; Klabunowski 1957, 1959). This research group studied the reaction of 2-methylhexanone with cyanoethylene in the presence of quartz doped with alkali ions. Besides a low efficiency of such reactions, it is mere speculation that quartz crystal had an influence on the course of the stereochemical course of the evolution so long as any evidence for an excess of one sort of chiral quartz crystals is missing.

More successful and convincing were experiments with optically active α-methyl amino acids as components of catalysts of sugar syntheses via the Formose reaction (Pizzarello and Weber 2004, 2010; Breslow and Levine 2006). Partially enantiomer-enriched α-methyl amino acids were isolated from meteorites as discussed in Chap. 9, and thus were, in principle available on the early Earth, but in extremely low quantities. However, even when 100% optically pure α-alkyl amino acids were used as ligands of chiral catalysts, only 4.8% of erythrose and 10.6% of threose were obtained (Pizzarello and Weber 2004). Analogous experiment dealing with the enantioselective synthesis of pentoses was reported (Pizzarello and Weber 2010).

Ad (b) The change of a world originally consisting of racemates to a world dominated by L-amino acids and D-sugars may, in principle, also be explained by an enantioselective destruction of D-amino acids and L-sugars. For such a scenario, enantioselective catalysts might again have played a decisive role. First experiments in this direction were again performed with chiral catalysts based on doped, whereupon copper, nickel, and platinum ions played the role of active centers. When gaseous 2-butanol was blown over these activated quartz crystals temperature in the range of 40–500 °C, stereoselective dehydrogenation was detected to a remarkable extent despite the high temperatures (Schwab et al. 1912, 1934). Similar experiments were later reported by Klabunowski et al., but without focus on biomonomers (Terent'ev and Klabunowsku 1953).

To research groups tried to find out to what extent the rate of hydrolytic degradation of oligopeptides and polypeptides depends on their stereosequence (Brack and Spach 1980; Bleir et al. 1980). It was found for both, α-helices and ß-sheet structure forming prAAs it was found that random stereosequences suffer a faster degradation than homochiral peptides, because the secondary structures of homochiral polypeptides are thermodynamically and chemically more stable. In combination with a more efficient homochiral coupling in oligo and polypeptide syntheses, as it is discussed below, cycles of repeated polymerization and degradation reactions might have resulted in a process of increasing enantioselectivity and homochirality. Yet, such a cyclic process might have had a chance against competing racemization only at temperatures below 25 °C (s. Sect. 10.2).

A quite different approach was described by Garay in 1968, who exposed racemic Tyr to a spin-polarized ß-irradiation generated by the radio-active decay of ^{90}Sr. Yet, those experiments showed an extremely low efficiency even after 18 months, and thus, did certainly not have a chance to outpace racemization (Garay 1968). Furthermore, other authors were not able to reproduce those results (Bonner et al. 1975). A slight enantioselective degradation was observed for D,L-Leu in a shorter time, when it was irradiated with spin-polarized electrons (Bonner et al. 1975, 1975b; Bonner 1984). Further experiments with spin-polarized electrons, which might have existed on the early Earth, were later also reported by Hegstrom et al. (1985), but all those experiments have in common that their efficiency is extremely low. This conclusion is also valid for analogous experiments with spin-polarized positrons (Gidley et al. 1982).

Beginning with Van't Hoff (1887), several authors have speculated that circularly polarized UV light might have caused an enantioselective degradation of racemates. First successful experiments with circularly polarized UV light were reported by Kuhn and coworkers as early as in 1929 (Kuhn and Braun 1929, 1930), who used racemic α-bromo- or α-azido-propionic acid as substrates. First experiments with prAAs and circularly polarized laser light (wavelenth212.8 nm) were described by Flores et al. in 1977 and also by Norden in the same year (Flores et al. 1977; Norden 1977). Yet, the best results obtained with D, L-Leu was an enantiomeric excess of only 2.0–2.5%. More recently, detailed studies were performed by Meierhenrich and coworkers (Meierhenrich et al. 2005, 2010, 2011; DeMarcellus et al. 2011; Meinert et al. 2011, 2014, 2015, 2020; Modica et al. 2014). Those authors used circularly polarized synchrotron radiation with variation of wavelength. At first only a low "ee" of 1.2% was obtained for D,L-Ala, but in later experiments, the "ee" was enhanced up to 4.2% (Meinert et al. 2014, 2015). In the case of racemic Val, Norval, α-Abu, and 2,3-diamino-propionic acid, a maximum "ee" of 2.5% was achieved and for D,L-Leu a record of 5.0%. Those authors published the following interpretation of their results:

Amino acids asymmetry might have been induced by circularly polarized electromagnetic radiation that has been detected in interstellar environments and star-forming regions. These data and indications point toward the fact that biomolecular asymmetry might have been induced photochemically in interstellar amino acids before these amino acids were delivered to the early Earth and triggered the origin of life on our planet.

The following arguments are in contradiction to this interpretation.

1. The formation of prAAs with significant "ee" in interstellar ice is mere speculation based on experiments in laboratories.
2. It is questionable that a significant amount of optically active prAAs have been delivered to the early Earth.
3. All the prAAs from meteorites were racemic.
4. All the energies that were capable to produce prebiotic syntheses of prAAs (s. Sect. 10.4) were also capable to destroy prAAs, and in this way prAAs that rained down from space were also destroyed.
5. The rapid racemization of prAAs (s. Sect. 10.3) was not taken into account.

Ad (c) Numerous experiments were conducted, and new methods were developed to achieve a selective enrichment of one enantiomer either via enantioselective crystallization (Klabunowski 1959) or via selective adsorption on chiral surfaces. For example, Harada (1965) observed that one enantiomer preferentially crystallized from a solution, when a few crystals of this enantiomer were added to stimulate nucleation of the crystal growth. An "ee" around 90% was achieved in favorable cases. More recently, such experiments were resumed by R. Breslow (Breslow and Levine 2006; Breslow and Cheng 2009). For those experiments, an inverse approach was used. The individual enantiomers had a higher solubility than the racemates. Upon cooling of solutions containing a slight excess of one enantiomer, the racemates crystallized first and the "ee" of the remaining solution increased. A further enhancement of the "ee" was then achieved by repetition of the experiments with the "ee" enriched fraction. In this way, an "ee" enrichment up to 90% was achieved for Phe, uridine, and adenosine. Similar experiments were also reported by Hayashi et al. in 2006 (Hayashi et al. 2006). However, it should be emphasized that such fractionated crystallizations do never yield an increase of the absolute quantity of one enantiomer, they just enhance its concentration in a smaller volume.

Another highly efficient recrystallization method was developed for Asp by C. Viedna et al. (2008). In contrast to other prAAs, Asp can form separate crystals of D-Asp and L-Asp, when both enantiomers crystallize from the same solution. When different quantities of D- and L-Asp are heated in the presence of a solution which enables racemization, the mass of the crystallized predominant enantiomer increases steadily at the expense of the enantiomer. In this way, complete transformation of one enantiomer into the other one can be achieved. Unfortunately, this excellent result does not help to explain the emergence of homochirality in the chemical evolution. First, the crystalline enantiomer cannot react, for example, by polymerization. Second, this method only works in the extremely rare cases where the enantiomers crystallize in separate crystals from the same solution.

Numerous experiments based on association or adsorption were performed in such a way that the racemates were combined with optically active natural products such aa strychnine, brucine, cellulose, silk, or wool. Such experiments have a long tradition and trace back to experiments of the Nobel Prize laureate R. Willstätter, who used silk and wool as optically active reaction partners (1904). Yet, such experiments are not helpful for establishing the hypothesis of chemical evolution, because they require the existence of homochiral biopolymers. However, they suggest that adsorption of biomonomers on preformed biopolymers prior to their reaction might have had a reinforcing effect for an enantioselectivity and homochirality favoring polymerization process. Enantioselective adsorption on optically active synthetic polymers is a working field which has attracted much interest over the past 50 years. The main stimulus came from the intention to develop an analytical approach allowing for the separation of enantiomers of natural or synthetic compounds. This goal has satisfactorily been achieved in modern versions of gas or liquid chromatography (Lindner 1992). The second method has the important advantage that the volatility of substrate does not play any role.

With regard to enantioselective adsorption on the early Earth, the role of alumino-silicates is of particular interest, because such minerals were present on the Earth crust at any time and everywhere. Achiral alumino-silicates can, of course, not distinguish between enantiomers, but they can differentiate between diastereomers. A slight extent of stereoselective adsorption of mandelic acid or lysergic acid esters has, indeed, been described pretty early by Stoll and Hoffmann (1938) and by Jamison and Turner (1942). A considerable extent of enantioselective adsorption of prAAs was found on chiral quartz surfaces (Bonner and Kavasmaneck 1967; Bonner et al. 1975, 1975b; Kavasmanek and Bonner 1977). The extent of the enantioselectivity depends very much on the structure of substrate. It may be high for prAAs and saccharides (Holzapfel 1951), but it is low for non-polar compounds. A particularly interesting phenomenon is the enantioselective adsorption of chiral metal complexes of chromium, cobalt, or platinum in quartz (Tsuchida and Kobayashi 1936; Schweitzer and Tanlbot 1952; Kuebler and Baihar 1953; Balandin 1955). This phenomenon might have had the consequence that in this way catalysts for stereoselective syntheses of natural products were introduced into the chemical evolution.

In addition to quartz, also other minerals were studied as adsorbing matrix. In 2001, Hazen et al. reported on the selective adsorption of D- and L-prAAs on calcite (Hazen et al. 2001). Pretty amazing is a report of Bondy and Harrington (1979), who describe a highly selective adsorption of L-Leu, L-Asp, and D-glucose on colloidal bentonite. Bentonite mainly consists of montmorillonite with minor constituents of quartz, pyrite and calcite. Bondy and Harrington did not explain why they had selected bentonite for their experiments, in as much, as these minerals favor the racemization of prAAs (s. Sect. 10.3). Furthermore, those authors did not forward any explanation for a highly enantioselective matrix. The author has not found a reproduction of these experiments nor any other citation in the literature dealing with chemical evolution.

Well understandable, however, is the observation of Bujdak et al. (2006), who studied the adsorption of diastereomeric dipeptides on the surface of minerals and found a preferential adsorption of the homochiral dipeptides. Since diastereomers possess different conformations and energy contents, a stereoselective adsorption is not surprising. When a selective adsorption on the surface of minerals also favored homochiral coupling (mere speculation), the adsorption might have made a significant contribution to the growth of homochirality in the course of the evaluation. In this connection, it is of interest to note that lower temperatures should have dual positive effect, because they have favored adsorption and reduced the negative influence of racemization (s. Sect. 10.3). At this point, the enantioselective adsorption of oligopeptide blocks (15 or 17 L-AA units) onto a 32 L-units long preformed polypeptide should be mentioned (Saghatelian et al. 2001). The analogous polypeptide made up by D-AAs was not adsorbed. However, the basic problem, how long (32 units) blocks of L-units were formed in the early period of the chemical evolution is hard to explain.

Experiments designed to produce a large "ee" via syntheses involving optically active catalysts were mainly performed with regard to syntheses of saccharides via the formose reaction. When Pizzarello and Weber (2004) studied the condensation of glycolaldehyde with addition of 100% optically active R- or S-α-methyl isovaleric acid, they obtained erythrose or threose with an "EE" up to 50%. This result is of particular interest, because the chiral cocatalyst is insensitive to racemization under alkaline conditions and because this amino acid was isolated with a moderate "ee" from meteorites (s. Chap. 9). Hence, those authors hypothesized that chiral (co)catalysts delivered from space might have spurred the stereoselective synthesis of D-saccharides, incl. D-ribose, on the early earth. This is certainly an interesting and, at first glance reasonable, hypothesis, but it has several weak points. First, the amount of optically active α-methyl AAs was certainly extremely low. Second, a satisfactory synthesis of D-ribose under the influence of the chiral α-methyl AA was not proven. Third, D-ribose is a rather unstable molecule, and it is absolutely unlikely that D-ribose was accumulated in a primordial soup over a period of many millions of years.

Pizzarello and Weber extended their studies of saccharide syntheses to the enantioselective preparation of pentoses starting out from glycolaldehyde and glycerol aldehyde. In contrast to their work on tetroses, those authors used in this case dipeptides of normal L-prAAs. Using L-Val-L-Val or L-Ile-L-Val, moderate yields of mixtures of D-ribofuranoside and D-ribopyranoside were obtained having an "ee" up to 44%. For other saccharides extracted from the reaction mixture, the "ee" varied over a broad range depending on the chemical structure. Similar results were also published by other research groups. For example, Cordova et al. (2005) and Casas et al. (2005) demonstrated for polycondensations of O-Bn-glycolaldehyde that the "ee" increased linearly with the amount of L-Pro added as cocatalyst. Breslow and Cheng (2010) or Hein et al. (2011) also reported a significant increase of the "ee" for syntheses of glycerol aldehyde or other saccharide precursors from glycolaldehyde when L-prAAs were added as cocatalysts. It is obvious that stereoselective syntheses of saccharides or prAAs via condensation reactions involving optically active catalysts or cocatalysts are a particularly efficient strategy to achieve homochirality.

An unexpected and interesting process allowing for an unusually efficient enrichment of one enantiomer was reported by Soai and coworkers (Kawasaki et al. 2005, 2006, 2008; Mineki et al. 2012). This research group produced chiral crystals from achiral molecules, such as hippuric acid, adenine, or cytosine, by crystallization from stirred solutions. These chiral crystals were, in turn, used as cocatalyst in an autocatalytic synthesis which yielded an "ee" up to 90%. This method is, one the one hand, an important example for the success of an autocatalytic synthetic process starting out from achiral organic compounds, but, on the other hand, this approach is limited to the transformation of pyridine-2-carbaldehyde into pyridine-2-butanol so far. As long as no analogous autocatalytic synthesis of D-saccharides or L-prAAs has been found, it is not justified to consider the results of the Soai group as example for the enantioselective evolution of homochiral biopolymers.

10.2 Enantio-Selection Via Homochiral Coupling

This section is dedicated to the question, if and how a preferential homochiral coupling of biomonomers might have contributed to the emergence of homochirality in the course of chemical evolution. The most intensively studied group of reactions are syntheses of dipeptides, oligopeptides, and polypeptides. The speculation undelaying this working field is the assumption that a local excess of one enantiomer was reinforced by predominance of homochiral coupling in combination with faster degradation of products resulting from heterochiral coupling. For enumeration and evaluation of the pertinent experiments, it is advisable to differentiate between (poly)condensation and condensative chain polymerization (CCP, s. Chap. 3). A predominance of homochiral coupling in a polycondensation was first reported by Degens et al. (1970). When those authors heated D-, D,L- or L-aspartic acid on kaolin, they observed that the polycondensation of the D-enantiomer was slower than condensations of the other stereoisomers. Unfortunately, the chemical structure of the reaction products was not characterized that the resulting polycondensates were free of ß-amide and imide groups. Polycondensation experiments in solution were conducted by Brack and Spach (1971a, 1971b), Spach (1974), using the pentachlorophenyl esters of γ-O-Bn-D-Glu and γ-O-Bn-L-Glu as monomers. Preformed polyglutamates having different sequences of D- and L-units were used as macro-initiators. From kinetic measurements, those authors deduced that the observed predominance of homochiral coupling was mainly a consequence of the α-helix structure and not a result of the D- or L-unit forming the helix chain end. However, for the following reasons, this conclusion is not justified. First, intensive racemization occurred under the given reaction conditions. Second, the formation of cyclic dipeptides was ignored. Third, those authors did not understand the difference between polycondensation and CCP and the consequences for the interpretation of their results. Fourth, neither the molecular weights of the reaction products nor the helix fraction was determined.

Extensive studies of (poly)peptide syntheses including a quantitative evaluation of the coupling reactions were performed by Kricheldorf et al. (Kricheldorf and

Scheme 10.1 Formation of enantiomeric and diastreomeric dipeptides in the condensation of racemic amino acid derivatives

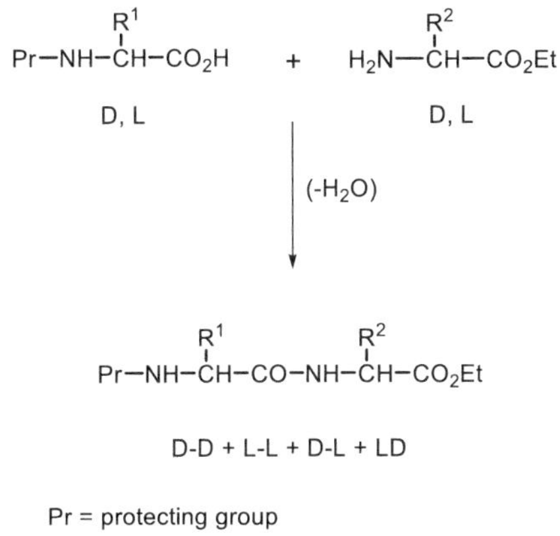

Mang 1982a, 1983; Kricheldorf and Au 1985; Kricheldorf et al. 1985). The majority of the condensation experiments were syntheses of dipeptides from racemic N-protected prAAs and racemic prAA alkyl esters. These condensations yield mixtures of stereoisomeric dipeptides (Scheme 10.1) that were analyzed by ^{13}C- and ^{15}N MMR spectroscopy. These spectroscopic methods don't allow for a distinction between enantiomers, but they allow for a differentiation and quantification of diastereomers. In other words, the molar ratio of L-L (+ D-D) peptides versus D-L (+L-D) peptides can be determined and thus the extent of homochiral coupling. Starting out from racemic Ala, leu, Phe, and Val derivatives, 46 condensations were studied based on coupling with 2-ethoxy-1-ethoxycarbonal-1,2-dihydroquinoline (EEDQ). In 40 experiments, a predominance of homochiral coupling was found with L-L/L-D ratios in the range of 1.2–1.4. On the other hand, evaluation of 15 dipeptide syntheses promoted with carbonyldiimidazole (CDI) yielded L-L/L-D ratios only in the range of 0.8–1.0. Furthermore, diastereomeric tripeptides were prepared from racemic N-protected dipeptides and racemic prAA alkyl esters by means of EEDQ as coupling agent. In this series, 12 out of 18 experiments showed a prevalence of homochiral coupling in the range of 1.1–1.8. In other words, the success of predominant homochiral coupling amounted to 75% in the case of tripeptides and to more than 80% for dipeptides.

Numerous oligopeptide syntheses were conducted with coupling agents that might have existed on the early Earth (in contrast to EEDQ). For example, cyanoacetylene is known to be a component of interstellar clouds. In seven out of 12 dipeptide syntheses, L-L/L-D ratios >1.0 were found with values up to 4.4 (!). With *N,N′*-bisalkylcarbodiimides or with (bisalkyl) cyanamides, the following results were obtained. All seven experiments performed with cyanamide in water yielded predominantly homochiral dipeptides with L-L/L-D ratios up to 1.9. When diethyl cyanamide was used as condensing agent in organic solvents, nine out of 12 experiments yielded L-L/L-D ratios above unity with a maximum value of 3.9. Nine out of 10 dipeptide syntheses performed with diethyl cyanamide in bulk showed a predominance of homochiral coupling. When diisopropylcarbodiimide was used as coupling agent in water, all seven experiments were successful and yielded L-L/L-D ratios up to 4.0. It is noteworthy, that the highest L-L/L-D ratios in all series of oligopeptide syntheses were achieved when D.L-Val derivatives were involved, although Val is not a helix forming prAA. Apparently, the steric demands of the branched side chain are decisive for the stereoselectivity of the coupling process.

Additional five dipeptide syntheses were conducted in water with a water-soluble carbodiimide. All five experiments favored homochiral coupling. Furthermore, six out of 10 coupling experiments conducted with *N,N′*-dicyclohexylcarbodiimide in organic solvents yielded L-L/L-D ratios >1.0 with a maximum at 3.9. A large number of dipeptide and tripeptide syntheses were conducted with various phosphoric acid derivatives. With poly(ethyl)phosphate as condensing agent, all 32 experiments yielded a predominance of homochirality with L-L/L-D ratios up to 3.0. All 14 analogous tripeptide syntheses also favored homochiral coupling. With poly(ethyl) phosphate and three other phosphorous condensing agents, it was also feasible to perform dipeptide syntheses with high yields in organic solvents when high concentrations of all reagents were used (Kricheldorf and Mang 1983). All 20 experiments with D,L-Leu derivatives favored homochiral coupling, but only 50% of the 18 experiments with racemic Val compounds favored homochiral dipeptides. The other 50% of these D,L-Val experiments yielded L-L/L-D ratios around 1.0. A predominance of homochiral coupling was also observed for D- and Lys derivatives in monolayers on the surface of water (Zepiki et al. 2002).

In summary, it may be said that nearly 80% of the 230 oligopeptide syntheses performed with racemic prAA derivatives favored homochiral coupling, and when the experiments with carbonyldiimidazole are excluded the success of homochiral coupling increases to approximately 90%. An exclusion of CDI makes sense, because it cannot have played a role for the formation of biopolymers due to its high sensitivity to water and ammonia. Furthermore, CDI has not been identified in interstellar clouds or in meteorites. A predominance of homochiral coupling on the order of 90% is particularly noteworthy considering that this result was not influenced by the presence of helical reaction partners.

The extent of homochiral coupling was also examined for condensative chain polymerizations (CCPs) of racemic cyclic derivatives of prAAs such as NCAs, TOOs, and TADs (Scheme 10.2). E.R. Blout and P. Doty were the first to study the kinetics of the CCPs of D-, L- and D, L-g-O-Bn-Glu-NCA (Blout and Idelson 1956; Blout et al. 1957; Lundberg and Doty 1957; Idelson and Blout 1958). They found that the pure D- and L-enantiomers polymerize with identical rates, whereas the polymerization of the racemate was significantly slower. Those authors concluded that the helical secondary structure of stereohomopolymers favors the homochiral chain growth, whereas the assumed random stereosequence destabilizes the helical secondary structure and thus in turn, the polymerization rate. In the following decades, several research groups continued to study the stereochemical course of the NCAs derived from glutamates, leucine, and valine (Matsura et al. 1965; Inoue et al. 1968; Tsuruta et al. 1967; Akaike et al. 1975). Those studies were followed by a paper of F.D. Williams (Williams et al. 1971) and by several publications of Imanishi and Higashimura (Imanishi et al. 1973, 1974, 1977; Hshimoto et al. 1976). Further studies in this field were published by W. A. Bonner (Bleir et al. 1980;

Scheme 10.2 Various polymerizable cyclic anhydrides of α-amino acids

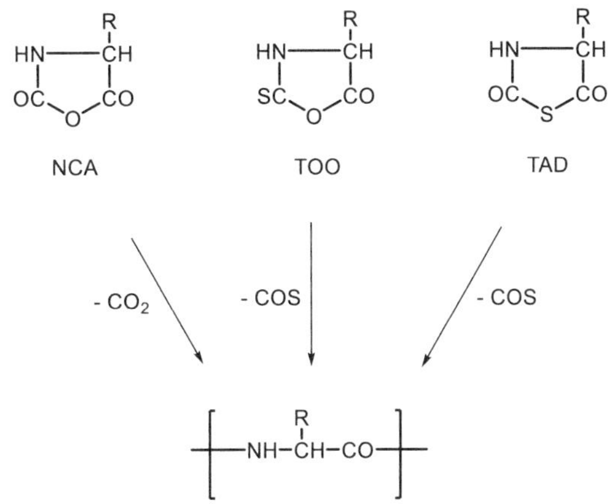

Bonner et al. 1981) and finally a study of trypsin-NCA is worth mentioning (Blöcher et al. 2001).

All these research performed kinetic studies of monomer mixtures containing either more L- than D-NCAs or vice versa. The polymerizations were stopped after a certain time with a variation of the conversion, and the composition of the isolated polypeptides was analyzed. For the α-helix forming prAAs, Ala, Glu, and Leu, a preferred incorporation of enantiomer used in excess was found, so that mainly homochiral sequences were formed. In the case of the ß-sheet forming valine another trend was observed. Either there was no enantioselectivity detectable or heterochiral sequences were slightly favored (Akaike et al. 1975; Bonner et al. 1981). This stereochemical difference was already predicted by Wald in 1957.

Numerous polymerizations of D,L-Ala-NCA, D,L Leu-NCA, D,L-g-O-Bn-Glu--NCA and D,L-Val-NCA were studied by Kricheldorf and coworkers using 13C- or 15-N NMR spectroscopy as analytical method (Kricheldorf and Hull 1979, 1982; Kricheldorf and Mang 1981, 1982a, 1982b). With D, L-Ala NCA, two out of six polymerizations favored homochiral chain growth, but two preferred heterochiral coupling and in the case of D,L-Val-NCA only two of five experiments showed homochiral chain growth, whereas three polymerization favored heterochiral coupling. However, all 10 polymerizations of D,L-Leu-NCA involved a preference of homochiral coupling and all six polymerizations of D,L-g-O-Bn-Glu-NCA likewise (Kricheldorf and Hull 1979, 1982). It is particularly interesting to see that the α-helix forming polypeptides favor homochiral sequences, because this finding perfectly agrees with the afore-cited kinetic studies of other research groups.

Whereas the ^{15}N NMR spectroscopy only allows for qualitative evaluation, the ^{13}C NMR spectra also enable a quantitative evaluation of diastereomeric dyads. When D,L-Leu-NCA or D,L-Val-NCA were polymerized in the melt at 120 °C, all resulting polypeptides consisted of random D/L sequences. However, when the polymerizations were conducted at 20 °C in solution in combination with protic initiator, all 12 experiments yielded a predominance of homochiral stereosequences with L-L/L-D ratios up to 2.0 (Kricheldorf and Mang 1981, 1982a, 1982b). In analogous polymerizations of D,L-Val-NCA, only 7 out of 12 experiments favored homochiral sequences and the L-L/L-D ratios were all below 1.3.

Surprisingly, the L-L/L-D ratios were higher, when the sulfur analogs of the NCAs (Scheme 10.2) were polymerized. All nine polypeptides that were obtained from polymerizations of D,L-Leu-TOO were mainly homochiral, and for six poly(D/L-Leu) samples, L-L/L-D ratios in the range of 4.0–4.5 were found. The six poly(D/L-Leu) samples prepared by polymerization of D,L-Leu-TAD were all predominantly homochiral with L-L/L-D ratios up to 4.1. The most amazing result was obtained from polymerizations of D,L-Val-TOO, because all 13 polymerizations favored homochiral chain growth with L-L/L-D ratios up to 3.1. For the sum of all these polymerization experiments, the conclusion may be drawn that approximately 75% of these polymerizations favored homochiral sequences. In the case of the α-helix forming leucine, this quotient is near 100%, when only polymerizations in solution below 50 °C were considered.

When all the more than 400 condensation and polymerization experiments are compared, it turns out that 75–85% of all experiments gave predominantly homochiral sequences and in most cases (oligopeptides) in the absence of helices. Hence, it may be concluded that the preference of homochiral coupling is somehow inscribed in the orbitals of the prAA derivatives used as monomers. The preferred homochiral coupling may be considered as a reinforcing mechanism for the formation of homochiral biopolymers starting out from a small excess of one enantiomer somewhere on the early Earth. The efficiency of such a process may be enhanced, when it was combined with a faster degradation (e.g., by hydrolysis or UV radiation) of less stable secondary structures rich in heterochiral dyads. At this point, the question comes up, if a reinforcing mechanism based on predominant homochiral coupling was efficient enough to play a decisive role for the evolution of homochiral peptides and proteins. The following counter-arguments have to be considered. First, the polymerization of prAA-NCAs cannot have played an important role for the evolution of polypeptides and proteins as discussed in Chap. 3. Second, there is no acceptable hypothesis (backed by experiments), which can explain how polypeptides were formed under prebiotic conditions outside the RNA world theory, which postulates that a RNA-based process yielding proteins with perfect control of homochirality preceded the formation of proteins. Third, it is hard to believe that the combination of a small local excess of one enantiomer and predominant homochiral coupling was more efficient than the influence of racemization and epimerization (s. Section below).

In this connection, a new publication of Weller (2024) entitled "The mystery of homochirality on Earth" needs discussion. In Chap. 5, that author claims "The mystery disappears" because of the chirality inherent in higher structures: "As shown with some real examples, a polymer of growing length will at some point spontaneously form a secondary and with longer chains a tertiary structure, which always are asymmetric" Weller ignores that the polypeptides of most prAAs form ß-sheet structures and not helices, and even the oligopeptides of helix forming prAAs need at least a length of 9–15 repeat units (depending on structure and environment) before they adopt a helical conformation. The concept of Weller is based on the

"Protein-World" hypothesis meaning that a prebiotic evolution of proteins occurred before or parallel to the evolution of RNAs. He cites the syntheses of proteinoids as evidence for his concept ignoring all the negative arguments discussed in Chap. 6, and he cites the boric acid catalyzed synthesis of polyglycine (up to a DP of 30) reported by Sumie et al. (2023). The problems inherent in that approach were also discussed in Chap. 6.

The combination of preferential homochiral coupling (as demonstrated by the author) and its reinforcement by a helical secondary structure is certainly an effective mechanism supporting the emergence of a homochiral world, but it has two weak points, when the formation of homochiral proteins before the formation of RNAs is considered. First, the absence of any plausible explanation of a prebiotic synthesis of proteins (s. Chap. 6) and, second, the impact of racemization. As discussed below in more detail, partial racemization of prAAs requires a few years at 25 °C or a few months at high temperatures, whereas the evolution of proteins must have required many millions of years. This problem is less severe, when the evolution of proteins has occurred at temperatures below 0 °C according to the "Ice-World" theory of Trinks and Biebricher (s. Sect. 2.9), because the rate of racemization decreases by several orders of magnitude for temperatures below 0 °C. Yet, how were proteins formed at such low temperatures? The combination of homochiral coupling and formation of chiral helices as origin of the homochiral world we have today is more promising in the frame of the "RNA-World" hypothesis, because nucleotides and RNAs are far less sensitive to racemization than peptides at a slightly alkaline pH. Weller correctly states that even in a random condensation of enantiomeric monomers, a small percentage of homochiral oligomers may be formed. Homochiral oligonucleotides may have, in turn, aced as templates for the synthesis of a complementary strand via base pairing. After dissociation, this complementary oligonucleotide may itself have played the role of template. Once homochiral RNAs have developed a protein synthesizing machinery (what needs to be proven), they may have learned to synthesize proteins at a rate outpacing racemization in analogy to what happens in living organisms (s. below). This scenario is, of course, nothing but a fiction so long as experimental evidence is lacking. However, first supporting experiments were published by Joyce et al. in 1984 (Joyce et al. 1984). Those authors studied the oligomerization of D,L-guanosine-5-phosphatoimidazolide in the presence of poly(D-cytosine phosphate). They observed a high enantioselectivity in the presence of the template, but not in its absence. These experiments don't explain how the first homochiral RNAs were formed under prebiotic conditions, but they demonstrate that homochiral elongation of oligomeric RNAs and de novo syntheses of RNAs in the presence of a homochiral template were efficient enantioselective processes. The "Joyce experiments" are certainly a first step in the right direction, but they are still far from proving the afore-mentioned scenario of RNA-catalyzed formation of homochiral proteins. Furthermore, the existence of efficient homochiral chain growth does not explain, why and how L-sugars and D-amino acids have vanished. Hence, in the eyes of the author, the mystery of homochirality has not vanished yet.

10.3 About the Role of Racemization (Epimerization)

The terms racemization and epimerization summarize various chemical reactions that have in common that an enantiomer is transformed into its mirror image. The term epimerization concerns racemization of individual repeat units in a polymer, whereas racemization of a polymer means that all repeat units were transformed into the enantiomeric units. Complete racemization of a small molecule means that 50% of the original enantiomers were transformed into their mirror images, and the resulting product is then called racemate. In the case of chiral sp3-hybridized carbons, racemization requires that one substituent is temporarily removed from the carbon (usually a proton), so that a trigonal planar sp2-hybridized carbon is formed as a reaction intermediate. The temporarily removed substituent can now return from both sides of the planar intermediate with equal probability and either restores the original enantiomer or produces its mirror image. Since the author has not come across systematic racemization studies of nucleosides or nucleotides, the following text is focused on prAAs and polypeptides or proteins.

In the case of prAAs, the racemization or epimerization always proceeds via deprotonation of the α-proton. This process is favored by the following five factors:

(a) High temperatures. Table 10.2 (Bada 1971) illustrates that even the small increase of the temperature from 0 to 25 °C dramatically enhances the rate of racemization. The change from 0 to 100 °C accelerates the racemization by factor of 5,000,000 (!) for Asp and 7,000,000 in the case of Ile.
(b) Substituents at amino group, ß-carbon or carbonyl group that diminish the electron density, because such substituents enhance the acidity of the α-proton.
(c) Cyclic structures, such as cyclic oligopeptides, *N*-carboxyanhydrides of prAAs or prAAs as ligands of chelate complexes. Responsible for this "ring effect" is the entropy change resulting from the deprotonation. Deprotonation yields a trigonal planar carbanion with delocalization of the electric charge (s. Scheme 10.3 Eq. (a)), and this delocalization hinders rotation around the σ-bond connecting CO and α-C. The loss of rotational motion means loss of entropy, and this loss cannot happen in cyclic structures where rotation around the CO–α-C bond is impossible anyway.
(d) Basic catalysts or a basic environment as it may have existed in the presence of ammonia or in the presence of calcium carbonate.
(e) Acidic catalysts or acidic condition that can protonate the CO group of prAAs or peptides (s. Scheme 10.3 Eq. (b)). The acid catalyzed racemization of prAAs

Table 10.2 Half-lives of prAAs in aqueous solution at pH 7.5 (Bada 1971)

prAA	0 °C	25 °C	100 °C
Asp	430,000 y	3500 y	30 d
Phe	160,000 y	2000 y	–
Ala	1,400,000 y	12,000 y	120 d
Ile	6,000,000 y	48,000 y	300 d

Scheme 10.3
Mechanisms of (**a**) alkaline racemization and (**b**) acid-catalyzed racemization of α-α-amino acids

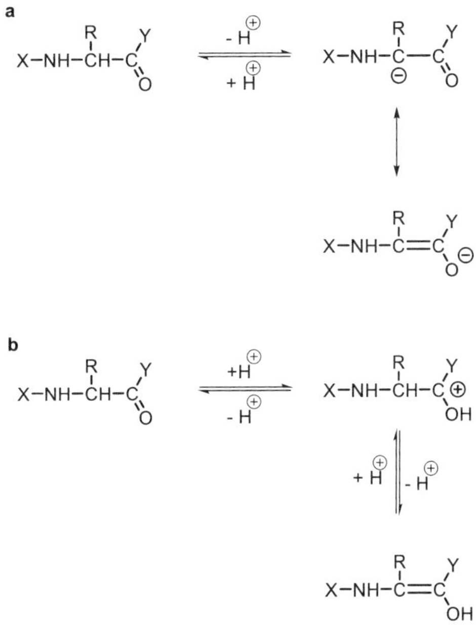

requires strong acids and will not have played a role for the chemical evolution in contrast to the base-catalyzed racemization.

The higher the temperature the lower must be the basicity of a catalyst which attacks the α-proton. At temperatures around 90–100 °C, as they exist in the neighborhood of black smokers or on the slopes of volcanoes, even water may play the role of the "basic" catalyst.

One may, of course, speculate that the high pressure on the ground of oceans reduces the rate of racemization. However, the influence of pressure on chemical reactions is based on a volume change in the course of the reaction, but a proton transfer requires almost not volume change. If the high pressure would reduce protonation/deprotonation reactions, it would hinder reactions of more space demanding groups even more. The difference between the rate of racemization and rates of other reactions would increase, and thus, be unfavorable for the emergence of homochirality in the neighborhood of hot vents.

Measurements concerning the racemization rates of prAAs or peptides were performed under various conditions (Bada 1972; Bada and Schroeder 1975; Bada and Miller 1987). A couple of particularly interesting results were summarized in Tables 10.2, 10.3, and 10.4. A comparison of racemization rates of three prAAs (Table 10.4) suggests that these rates reflect the influence of the substituents in ß-position. The alkyl groups of Ile have weak positive inductive effect, the CO group of Asp a negative effect. This interpretation also suggests that Ser and His belong to the relatively racemization sensitive prAAs. The tremendous influence of the temperature has already been mentioned above (point (a)). The comparison of half-live data

Table 10.3 Half-lives of three prAAs or their repeat units in proteins

	L-Aspartic a.	L-Alanine	L-Isoleucine
prAA:			
Free prAA	30 d	120 d	300 d
Asp-Phe-OM (aspartame)	13 h	-----	-----
Protein	1–3 d	-----	-----
In vivo:			
Mammalian teeth	350 a	-----	-----
Erythrocytes	3 a	-----	-----

Table 10.4 Comparison of racemization rates of free prAAs and prAA units in proteins

Reaction conditions	Isoleucine	Alanine	Aspartic a.
Free prAAs in H_2O at pH 7.6 and 135 °C	1.0	2.4	8.8
prAA units in insulin and in RNAs studied in H_2O at pH 7.6 and 135 °C	1.0	~3.0	8–9
prAA units in foraminifres and in deep sea drilling samples (age:210,000 years)	1.0	1.5	1.9

measured at 100 °C underlines this temperature effect. Furthermore, one has to take into account that the racemization rates increase by a factor of 100–1000, when prAAs play the role of ligand in a chelate complex. For example, the half-live of an Asp/Cu complex measured at 5 °C and pH 7.6 amounted to only 100 years (Bada and Man 1980). One might object that Cu ions were extremely rare in the early oceans, but Mg and Ca ions were certainly present in quantities similar to those known today, and in the neighborhood of the abundant FeS and NiS minerals, the corresponding metal ions were present. Hence, from the view point of complexation and temperature, the FeS chemistry in the neighborhood of black smokers favored by Wächtershäuser et al. (s. Chap. 5) is the least favorable birthplace of a chemical evolution.

The adsorption of growing biopolymer chains on silicates or their intercalation in layered silicates has been considered by several researchers as important prerequisite for a successful chain growth. Yet, alumo-silicates are polymeric anions which upon contact with peptides may accelerate their epimerization. Measurements confirming this assumption for montmorillonite have indeed been published by Frenkel and Heller-Kallai 1977. This finding is also one more argument against the "Silicate-World" hypothesis of Cairns-Smith. Furthermore, basic minerals such as magnesium and calcium carbonates accelerate the racemization of prAAs and peptides. Hence, the presence of basic minerals needs to be considered, when the age of protein-containing sediments (contributed by shell fish) is to be determined from the degree of racemization/epimerization (Bada and Schroeder 1975). Moreover, the rate of epimerization of peptides and proteins may be influenced by degradation reactions such as formation of cyclic dipeptides (2,5-diketopiperazines) via "back-biting" of the amino end group. The AA units of these cycles racemize by a factor 100–1000 faster than the same AA units in liner chains due to the entropy effect discussed above under point (c).

Epimerization of proteins does not only proceed in dead plants and animals, but also in living organisms (Ritz et al. 1996; Ritz-Timme and Collins 2002; Dobberstein et al. 2010; Truscott et al. 2016). If a protein after its synthesis is not subject to metabolism, its degree of epimerization may be sued as measure of the age of the organism under investigation. Due to the relatively high sensitivity of Asp to base-catalyzed racemization, this analytical approach is mainly focused on isolation and characterization of this acidic prAA. This approach is valid, for example, for proteins in dentin and for osteocalcin in bones. Decisive for the success of this analytical method is a perfect separation of the non-active proteins from those which are subject to a permanent metabolism (Ritz et al. 1996). When this analytical approach is perfectly performed by an expert, it allows for an accurate determination of the age of a corpse. Hence, this relatively new method has found increasing application forensic medicine.

Since racemization affects the secondary and tertiary structure proteins need for their biochemical and biological function, racemization can spur the outbreak of diseases and in extreme cases even become a mortal factor. Therefore, the biochemically active proteins have usually a short live time in all log living organisms. They are hydrolyzed and partially recycled and finally replaced by a freshly synthesized protein with 100% homochirality. In the case of microorganisms, the live span is short and the rate of proliferation is extremely high, so that a rapid substitution of damaged proteins is not necessary. Therefore, primitive organisms such as archea can survive even at high temperatures in the neighborhood of black smokers. The homochirality is a thermodynamically unfavorable state of biopolymers characterized by relatively low level of entropy, and racemization needs to be outpaced by an efficient kinetic process with permanent input of energy. Therefore, it is not justified to derive from the existence of primitive organisms at high temperatures that their environment was good birthplace of the chemical evolution. In the case of dead matter, the thermodynamic controls what happens and racemization generates a gain of entropy.

Finally, the existence of an indirect racemization process needs to be considered. Those energies which have enabled syntheses of prAAs from simple starting materials (s. Chap. 4) were also capable of destroying or modifying prAAs. Hence the concentrations of prAAs which existed in the primordial soup were the consequence of an equilibrium between syntheses and destruction or modification (e.g., methylation or hydroxylation via radicals), unless part of them escaped destruction by absorption in compartments. Now consider, a local excess of one enantiomer was formed on the surface of quartz or it was delivered from space (α-methyl AAs, s. Chap. 9), this excess would have been destroyed by degradation and formation of new racemic prAAs.

The sum of all these results and considerations and the high rates of racemization at pH >7 and temperatures >25 °C combines with the total racemization of prAAs in meteorites, and all together suggest that the emergence of homochiral proteins cannot have been the consequence of a time-consuming evolutionary process. This conclusion is valid so long as a mechanism which has blocked the racemization of prAAs has not been found. Absolute dryness prevents racemization (except via degradation), but chemical evolution cannot have been occurred in the total absence of water. Hence, Bada and Miller have published the following comment in 1987 (Bada and Miller 1987):

> Even if one or a combination of the above mechanisms produced chiral amino acids on the Earth, subsequent racemization would have rapidly taken place once again producing a racemic mixture. This point has been pointed out repeatedly (Miller and Orgel 1974; Keszthelyi et al. 1970; Dose 1981). …[…] Racemization rates in various proteins measured at elevated temperatures as well as in vivo racemization rates in living a mammalian proteins indicates that the rate of protein bound amino acids are faster than for free amino acids. This presents a potentially severe problem for the first organisms, especially since enzymes must have a high degree of [homo]chirality at the active side residues [….]. In order to preserve optically active amino acids in proteins, the protein synthesis rate would have to be rapid, and seems to be more likely that this was accomplished by biological rather than by abiotic processes. If this was the case, the origin of optically active amino acids in proteins may have only occurred after the process of biological protein synthesis was well established or else, it occurred simultaneously with the origin of this biosynthetic process. Thus, instead of the origin of optical activity amino acids preceding the origin of life on Earth, it may have occurred at the same time or after life on Earth became established.

The speculation that homochiral biopolymers might have been formed simultaneously with living cells or independently requires a strong belief in chemical miracles. With respect to their last sentence, Bada and Miller mentioned a concept of Spach (1984), who speculated that at first an achiral world of RNAs might have appeared which transformed itself in a chiral more efficient RNA world capable of synthesizing homochiral proteins. The nucleobase derivatives of glycerol C and D in Scheme 2.1 were proposed as examples of building blocks of achiral RNAs. Yet, this proposal is again nothing than wishful thinking without any experimental background. The author agrees with Bada and Miller in that considering the current state of knowledge (2023) the high rates of racemization of prAAs above 15 °C exclude that homochirality evolved as slow process in the course of many millions of years. Whatever the mankind will learn in the course of the coming decades and century, it will remain an undeniable fact that prAAs racemize quite rapidly in living organisms, when the time needed for 50% racemization is compared with the many millions of years required for the chemical evolution.

Finally, it should be mentioned that in almost all publications dealing with formation or enhancement of an enantiomer excess under prebiotic conditions, the term racemization is not even mentioned and the work of Bada is not cited. Even in books and reviews that appeared after 2000, the problem of racemization is not discussed.

10.4 The Physiological pH

At this point, the question needs discussion, why the physiological pH of humans and mammalians amounts to 7.4 and not to 7.0 or a slightly acidic value such as 6.5, At a pH of 7.4, the rate of racemization is approximately two orders of magnitude higher than at pH 7.0 and more than four orders of magnitude higher than at pH 6.5. In other words, the physiological PH is unfavorable for the emergence and maintenance of homochirality and the question is why.

Controlled by the brain-stem, the human organism is strongly concerned to maintain pH 7.4 within narrow limits (+/−0.05 pH units). This situation is most likely also valid for all mammalians, because the structure of their brain and the structure of the biopolymers making up body and brain are almost identical. The human and the mammalian organism is equipped with an efficient buffer system, the carbonate buffer, which enables to maintain the physiological pH even under extreme stress of the organism. Equation 10.1 represents the s. c. buffer equation. In the human organism, the acid HA is the carbonic acid H_2CO_3, which is according to eq. 10.2 in equilibrium with carbon dioxide and water. The basic component (A^-) is the hydrogen carbonate ion (the brackets indicate molar concentrations).

$$\mathrm{pH} = \mathrm{pK_a} - \log\left(\left[\mathrm{HA}\right]/\left[\mathrm{A^-}\right]\right) \tag{10.1}$$

with pK_a = dissociation constant of carbonic acid

$$\mathrm{H_2CO_3} \rightarrow \mathrm{CO_2} + \mathrm{H_2O} \tag{10.2}$$

Since even slight deviations from the physiological pH may have dramatic consequences for the activities of the human brain, it may be easily realized by self-experiment with hyperventilation. When more CO_2 is removed from the blood and brain than produced by combustion of glucose in the mitochondria, the pH tends to increase toward 7.5, 7.6, or even 7.7 and the person exercising hyperventilation gets a feeling of dizziness. Hyperventilation enforced against the "will" of the brain-stem might have lethal consequences.

This scenario leads to the question, whether the early ocean already had a pH around 7.4 and the first cells simply accepted and bequeathed this pH to the following generations of organisms without ambition to alter it. Or, the pH of the prebiotic ocean was different from pH 7.4 and the numerous generations of cells and organisms following the first cells developed and established pH 7.4 as consequence of an optimization process.

A definite answer on this question cannot be given, in as much as the pH of the prebiotic ocean is and will remain unknown. The oceans today have a pH around 7.5 slight varying with presence or absence of basic minerals (e.g., $CaCO_3$). However, the early oceans might have had a neutral or slightly acidic pH due to intensive volcanic activities. Typical for volcanic exhalations are gases such as CO_2 and H_2S which might have been oxidized to SO_2 which yields H_2SO_3 in contact with water. Furthermore, extremophilic microorganism (s.c. prokaryotes such as archaea) have been found in the neighborhood of acidic or basic hydrothermal vents which can survive. Species were found which can exist in alkaline environment with pH values up to 10, whereas other species can exist at pH values around 1. A group of *Thermoplasmatales*, the *picrophilus* reach an optimum growth rate even at pH 0.7! The mitochondria which produce ATP in every cell of living organisms are successors of archaea, They have developed a proton pump which allows for a broad variation of the pH inside and outside their cell membrane. All these observations demonstrate that cells can vary the pH over a broad range and can easily adopt their metabolism to various pH depending on their environment. Hence, it may be concluded that the higher organisms which appeared later in the biological evolution have optimized their internal pH according to their needs. If so, the question arises why a pH around 7.4 is an optimum pH for almost all higher organisms, although it is unfavorable from the view point of racemization.

The research results of Benner and Hutter (2002) provide an interesting answer to this question. Those authors synthesized structural analogs of oligoRNAs and varied the groups connecting the nucleosides. Instead of phosphate groups, for examples, sulfate or methyl phosphonate groups were incorporated. In this way, the anionic character which is typical for the RNA backbone at pH values above 7.0 had disappeared. Those authors drew the following conclusion:

The polyanionic nature of the backbone appears to be critical to prevent the single strands from folding them to act as templates, guiding the interaction between two strands to form a double helix in a way that permits simple rules to guide the molecular recognition event, and buttering the sensitivity of the physiochemical properties to change in sequence. We argue that the feature of polyelectrolyte (polyanion or polycation) may be required for a "self-sustaining chemical system capable of Darwinian evolution." The polyelectrolyte structure, therefore, may be an universal signature of life, regardless of its genesis and unique to living forms as well.

In addition to the afore-mentioned argumentation, a second argument can be forwarded, which supports the hypothesis that the polyanionic character of the RNA chains at a pH 7.0 is advantageous for existence and function of RNAs. This second argument is a higher stability against hydrolysis. Nucleophiles in general (incl. water and OH anions) are species having a high electron density and thus they are repelled by an anionic polymer or surface. In this regard, a study of Usher and Orgel (1976) is of interest. Those authors synthesized an oligo(A) dodecamer with 2,5′-connection in the middle of the chain and compared its hydrolytic stability in comparison with a dodecamer exclusively mad up by 3′5′-connections. In the presence of a poly(U) matrix, the 2′,5-analog was hydrolyzed more rapidly by a factor of 900. Hence, it seems that conformation and hydrolytic stability of RNA chains in the anionic form is a greater advantage for the survival of living organisms than a somewhat enhanced stability of proteins against racemization in neutral or slightly acid water.

Finally, it should be mentioned again that living cells are capable to compensate for damage of proteins by racemization or any other destructing event by rapid de novo synthesis of a 100% perfect protein. For proteins in a dead environment, this substitution mechanism does not exist and racemization takes its course dictated by a gain in entropy. Now consider an undirected chemical evolution and emergence of the first cells by chance events or contingency. How was the homochirality maintained without the metabolic activities of living organisms and how did the first cell suddenly learn that substitution/repair of damaged proteins is crucial for the survival of their species. The author has not found a discussion of this fundamental problem in publications advancing the chance-event or contingency concept.

References

Akaike T, Aoyaki T, Inoue S (1975) Biopolymers 14:2577
Bada J (1971) Adv Chem Ser 106:109
Bada J (1972) Chem Commun 94:1371
Bada J, Man EH (1980) Earth Sci Rev 16:21
Bada J, Miller SL (1987) Biopolymers 20:21 (and cited literature)
Bada J, Schroeder RA (1975) Naturwissenschaften 62:71 (and cited literature)
Balandin A (1955) Iz Akad Nauk USSR r 4:624
Bauer E (2012) Chem Soc Rev 41:3153
Benner S, Hutter D (2002) Bioorg Chem 30:62
Berkessel A, Groyer H (2005) Asymmetric Organocatalysis. Wiley VCH, Weinheim
Bleir N, Dirbas F, Bonner W (1980) Tetrahedron 3:27
Blöcher M, Hitz T, Luisi PL (2001) Helv Chim Acta 84:842
Blout E, Doty F, Young J (1957) J AmChem Soc 79:749
Blout E, Idelson M (1956) J Am Chem Soc 78:3857
Bondy S, Harrington M (1979) Science 203:1243
Bonner W (1984) Orig Life 14:285
Bonner W, Dart MA, Yearian MR (1975) Nature 258:419
Bonner W, Kavasmaneck PR (1967) J Org Chem 41:2225
Bonner W, Kavasmaneck PR, Martin FS, Flores JJ (1975b) Orig Life 6:367–376
Bonner WA, Blair NE, Dirbas FM (1981) Orig Life 11:119
Brack A, Spach G (1971a) Nat Phys Sci 229:124
Brack A, Spach G (1971b) Bull Soc Chim France 1971:4481–4485
Brack A, Spach G (1980) J Mol Evol 15:231
Breslow R, Cheng Z-L (2009) Proc Natl Acad Sci USA 106:9144
Breslow R, Cheng Z-L (2010) Proc Natl Acad Sci USA 107:5723
Breslow R, Levine MS (2006) Proc Natl Acad Sci USA 103:12979
Bujdak J, Remko M, Rohde B (2006) J Colloid Interface Sci 29:301
Casas J, Engqvist M, Ibrahem I, Kaynak B, Cordova A (2005) Angew Chem Int Ed 44:1343
Cordova A, Enquist M, Ibrahem I, Sunden H (2005) Chem Commun 2005:2047
Degens E, Matheia J, Jackson TA (1970) Nature 227:492
DeMarcellus P, Meinert C, Nuevo M, Flipi J, Danger G, Deboffle D, Nahon L, Le Sergeant d'Hendecourt L, Meierhenrich U (2011) Astrophys J Lett 727:L27
Dobberstein R, Tung SM, Ritz-Timme S (2010) Int J Legal Med 124:269
Dose K (1981) Orig Life 11:165
Faber K (2011) Biotransformation in organic chemistry – a text book. 6th edn. Springer, Berlin
Faber K, Griengel H (1992) Chirality in organic synthesis – The use of biocatalysts. In: Janoschek R (ed) Chirality from weak bosons to the α-helix. Springer, New York
Flores J, Bonner W, Massey G (1977) J Am Chem Soc 99:3622
Frenkel L, Heller-Kallai L (1977) Chem Geol 19:161
Garay S (1968) Nature 219:328
Gidley W, Rich A, VanHouse J, Zitzewitz PW (1982) Nature 297:639
Glorius F, Gnas Y (2006) Synthesis 2006(1):1899
Harada K (1965) Bull Soc Chem Jpn 38:1559
Hayashi Y, Matsuzawa M, Yamaguchi J, Yonehara S, Matsumoto Y, Shoji M, Haszime D, Koshino H (2006) Angew Chem Int Ed 45:4553
Hazen R, Filey TR, Goodfriend G (2001) Proc Natl Acad Sci USA 98:5487
Hegstrom R, Rich A, VanHouse J (1985) Nature 313:391
Hein J, Tse E, Blackmond D (2011) Nature 3:703
Holzapfel L (1951) Z Elektrochem 53:577
Hshimoto Y, Aoyama I, Imanishi Y, Hifashimura T (1976) Biopolymers 15:2407
Huber L, Trapp O (2022) Orig Life Evol Biosph 52:75–91

Idelson M, Blout ER (1958) J Am Chem Soc 80:2387
Imanishi Y, Aoyama A, Hashimoto Y, Higashimura T (1977) Biopolymers 16:187
Imanishi Y, Kugimya K, Higashimura T (1973) Biopolymers 12:2643
Imanishi Y, Kugimya K, Higashimura T (1974) Biopolymers 13:1205
Inoue S, Matsura K, Tsuruta T (1968) J Polym Sci Part C 23:721
Jacobsen N, Pfalz E, Escher T (1999) Comprehensive asymmetric catalysis. Springer, Berlin
Jakschitz T, Rohde B (2012) Chem Soc Rev 51:5484
Jamison M, Turner E (1942) Chem Soc 1942:611
Janoschek R (1992) Theories on the origin of biomolecular homochirality. In: Janoschek R (ed) Chirality – from weak bosons to the α-helix. Springer, New York
Joyce G, Vissert G, von Bockelt C, von Boom, Orgel L, van Westeren J (1984) Nature 310:662
Kavasmanek PR, Bonner WA (1977) J Am Chem Soc 99:44
Kawasaki T, Jo K, Igarashi H, Sato I, Nagano M, Koshima A, Soai K (2005) Angew Chem Int Ed 44:2774
Kawasaki T, Suzuki K, Hatase K, Otsuka M, Koshima K, Soai K (2006) Chem Commun 2006:1869
Kawasaki Z, Suzuki K, Hakoda Y, Soai K (2008) Angew Chem Int Ed 47:496
Keszthelyi L, Czege L, Fajszi C, Posfal J, Goldanski V (1970) Racemization and the origin of biomacromolecules. In: Walker DC (ed) Optical activity in nature. Elsevier, Amsterdam, pp 229–244
Klabunowski E (1957) Khimichevskaya Nauk I Premyshlenmost 2:197
Klabunowski EI (1959) In: Clark F, Synge RLM (eds) The origin of life on earth. Pergamon Press, New York
Kricheldorf H, Au M (1985) Macromol Chem Rapid Commun 6:469
Kricheldorf H, Au M, Mang T (1985) Int J Protein Res 26:149
Kricheldorf H, Hull W (1979) Maktomol Chem 180:1715
Kricheldorf H, Hull W (1982) Biopolymers 21:1635
Kricheldorf H, Mang T (1981) Makromol Chem 182:3027
Kricheldorf H, Mang T (1982a) Makromol Chem 183:2093
Kricheldorf H, Mang T (1982b) Makromol Chem 183:2113
Kricheldorf H, Mang T (1983) Int J Bioö Macromol 5:258
Kucbler I, Baihar J (1953) J Sam Chem Soc 75:4574
Kuhn W, Braun E (1929) Naturwissenschaften 17:227
Kuhn W, Braun E (1930) Naturwissenschaften 18:183
Latal R (1992) Parity violation in atomic physics. In: Janoschek R (ed) Chirality – from weak bosons to the α-helix. Springer, New York
Lindner W (1992) Strategies for liquid chromatographic resolutions of enantiomers. In: Janoschek R (ed) Chirality – from weak bosons to the α-helix. Springer, New York
Lundberg R, Doty P (1957) J Am Chem Soc 79:3961
Mason S, Tranter GB (1985) Proc R Soc (Lond) 397:45
Matsura K, Innoue S, Tsuruta T (1965) Makromol Chem 85:284
Meierhenrich U, Filipi J, Meiner C, Hoffmann S, Bredehöfft J, Nahon L (2010) Chem Biodivers 7:1651
Meierhenrich U, Filipi JJ, Meinert C, Hoffmann SV, Bredehöft JH, Nahon L (2011) In: Brückner H, Fuji N (eds) D-amino acids in chemistry in lifesciences and in biotechnology. Wiley VCH, Weinheim, pp 341–349
Meierhenrich U, Nahon L, Alcaraz C, Bredebhöft J, Hoffmann S, Barbier C, Brack A (2005) Angew Chem Int Ed 41:3630
Meinert C, Cassam-Chenai P, Jones NC, Nahon L, Hoffmann SV, Mweierhenrich UJ (2015) Orig Life Evol Biosph 43:149
Meinert C, deMarcellus P, LeSergent d'Hendecourt L, Nahon L, Jones NJ, Hoffnann SV, Bredehöft J, Meierhenrich U (2011) Phys Life Rev 8:307
Meinert C, Filipi J, Hoffmann SV, LeSergent d'Hendecourt L, deMarcellus P, Bredehöft J, Tgiemann W, Meierhentich UJ (2020) Symmetry 2:1055

Meinert C, Hoffmann S, Chassam, Chenai P, Evans AE, Girt C, Nahon L, Meierhenrich U (2014) Angew Int Ed 53:210
Miller S, Orgel S (1974) The origin of life on earth. Prentice hall Inc., Englewood Cliffs, pp 166–174
Mineki H, Hanasaki T, Matsumoto A, Kawasaki T, Soai K (2012) Chem Commun 48:10538
Modica P, Meinert C, deMarcellus P, Nahon L, Meierhenrich U, LeSergent d'Hendecourt L (2014) Astrophys J 788:79
Norden B (1977) Nature (London) 266:567
Pizzarello S, Weber A (2004) Science 303:1151
Pizzarello S, Weber A (2010) Orig Life Evol Biosph 40:3
Ritz S, Turzynski A, Schütz AW, Hoffmenn A, Rocholz G (1996) Forensic Sci Int 770:13
Ritz-Timme S, Collins MJ (2002) Ageing Res Rev 1:43
Saghatelian A, Yolobayashi Y, Soltani K, Reza-Ghadiri M (2001) Nature (let) 409:797
Schwab G, Rost F, Rudolf L (1912) naturwissenschaften 12:237
Schwab G, Rost F, Rudolf L (1934) Kolloid Z 68:157
Schweitzer G, Tanlbot C (1952) Chem Abstr 46:11004
Spach G (1974) Chimia 28:500
Spach G (1984) Orig Life 14:433
Stoll A, Hoffmann A (1938) Hoppe Seyler 251:155
Sumie Y, Sato K, KageKawa T, Furukawa Y (2023) Nature Commun Chem 6:89
Terent'ev A, Klabunowsku E (1953) Spornik Statel po obshehel Khimii 2:1521
Tranter G (1985) Nature (London) 318:172
Truscott R, Shey K, Friedrich M (2016) Trnds Biohem Sci 41:645 (Review)
Tsuchida R, Kobayashi M (1936) Bull Chem Soc Jpn 6:342
Tsuruta T, Inoue S, Matsuura K (1967) Biopolymers 5:313
Usher DA, Orgel LE (1976) Proc Natl Acad Sci USA 73:1149
Van't Hoff J (1887) The arrangement of atoms in space. 2nd edn, p 30. cited in Flores JJ et al. (1977)
Viedna C, Ortiz JE, DeTorres T, Izumi T, Blackmond D (2008) J Am Chem Soc 130:15274
Wächtershäuser G (1991) Med Hypotheses 36:307
Wald G (1957) Ann NY Acad Sci 69:352
Weller M (2024) Life (MDPI) 14:341
Williams F, Eshaque Y, Brown R (1971) Biopolymers 10:753
Willstätter R (1904) Ber Dtsch Chem Ges 37:3758
Winterfeld E (1992) Preparation of homochiral organic compounds. In: Janoschek R (ed) Chirality – from weak bosons to the α-helix. Springer, New York
Zepiki H, Shavit E, Tang M, Jensen T, Kjaer K, Bolbach G, Leiserowitz L, Weissbuch I, Lahav M (2002) Science 295:1266

Further Reading

Thiemann W (1974) Naturwissenschaften 61:470

Summary and Conclusions 11

Science [....] commits suicide when it adopts creed

T. H. Huxley

This chapter has mainly the purpose to summarize arguments that are in contradiction to the hypothesis of CE. This intention is not the consequence of a negative attitude of the author toward the potential existence of a CE, but it is thought as a counterbalance against the numerous overoptimistic and overinterpreting books and review articles that were published in the past. Furthermore, open questions and problems that have rarely or never been addressed should be formulated. The author has a neutral stance, and the words of Roger Bacon (Chap. 2) and the words of Thomas Huxley (s. above) are his guidelines.

11.1 Why This and Not That?

Only a small fraction of the numerous molecules that were detected in the model experiments or in the chemistry of space are present in the living organisms. This fact raises the question, why the bioactive compounds were selected in the course of the CE. A minor selection may, of course, also have taken place after the emergence of the first cell, but this consideration does not affect the importance of the aforementioned questions. Surprisingly, this question has rarely been asked and discussed in books and review articles dealing with the hypothesis of CE. Exceptions are the books of Wills and Bada (2000) and the book of Luisi (2006).

Luisi concentrates his discussion on the three most important classes of biomonomers: prAAs, saccharides, and nucleobases, and he justifies their selection with their high degree of thermodynamic stability. His original comment on p. 50:

H. R. Kricheldorf, *Life by Chemical Evolution?*
https://doi.org/10.1007/978-3-662-70752-4_11

> Going back to Miller's synthesis in the flask, one question is why α-amino acids have been obtained and not for example ß-amino acids, cyclic diketopiperazines and some other isomers. The answer is important: α-amino acids form, because they are the most stable products under the selected conditions. In other words, the formation of those amino acids is under thermodynamic control The same may be said for Oro's synthesis of adenine and other prebiotically low molecular weight substances formed in hypothetical vents, or found in space: certain molecules and no others form, because they are thermodynamically more stable.

This statement is wrong even in its first sentence. Large amounts of ß-alanine and smaller amounts of γ-amino acids were found in most model syntheses of amino acids and even more in the extracts of most carbonaceous meteorites. Furthermore, large amounts of sarcosine along with small amounts of other imino acids were detected in model syntheses and in extracts of meteorites. In one of the Miller experiments, described in his second publication in 1955, sarcosine was the main product (Miller 1955). Several prAAs such as Ser and Thr were only detected in traces by chromatographic analyses. It is far from correct to attribute prAAs a significantly higher thermodynamic stability over non-prAAs or imino acids. It is certainly true that almost all amino and imino acids are rather stable relative to many other classes of organic compounds (e.g., saccharides) due to their betaine structure. However, Wills and Bada also ignored that the reaction products of Millers experiments and the extracts of meteorites also contained large fractions of carboxylic acids.

The conclusion that the thermodynamic stability of prAAs has favored their selection and important role in the CE is not only one-sided and superficial it also raises other questions. A high thermostability has, for example, the consequence of a low reactivity. Activation or prAAs for peptides syntheses requires much energy in lab experiments and in living cells. A second problem connected with the selection of the prAAs is the question, why the known 20 prAAs were selected and not other α-amino acids. The author has never found any discussion of this problem in the literature about CE. The following three examples should illustrate this problem.

Example 1

Why was Ile selected in addition to Val (or vice versa), although both prAAs possess early identical properties, and although model syntheses and extracts from meteorites demonstrate that the probability of the formation of Ile is rather low. Furthermore, the influence of both prAAs on the secondary structure of polypeptides is almost identical. Both prAAs are hydrophobic and favor the ß-sheet structure. A similar problem is the selection of Thr in addition to Ser. Both are hydrophilic and favor the ß-sheet structure. The third problem of the same category is the selection of Glu in addition to Asp (or Gln in addition to Asn).

Example 2

Peptide bonds between prAAs and sarcosine have the property that the cisoid and transoid conformation have similar stabilities, what eases formation of loops, and sarcosine is a hydrophilic imino acid. Why was Sar not selected for loop-forming peptide sequences, above all in peptide sequences designed for a hydrophilic character?

Example 3

Ornithine is part of the urea metabolism and thus needed in the metabolism of almost all animals. Furthermore, Orn is a precursor in the biosynthesis of arginine. Why was Orn not used for the construction of proteins either alone or in combination with Lys? Model syntheses of amino acids and extracts of meteorites have shown that the formation of Lys is less likely than the formation of Orn.

The discovery of aptamers and their high selectivity for small molecules might bring about a solution of the afore-mentioned problems within the framework of the RNA world hypothesis. Consider aptamers were formed that were specific for prAAs and organized selection and transport of prAAs into compartments such as lipid vesicles containing a RNA machinery capable of synthesizing proteins. Yet, why have about 20 aptamers selective for prAAs emerged from a pool of hundred thousands of low molar mass RNAs? It is absolutely unlikely and, above all, unproven that only aptamers specific for prAAs can be formed, but no aptamers for non-prAAs. As long as a high predominance of prAA selective aptamers cannot be demonstrated in random syntheses of RNAs, the above-described mechanism for the selection of prAAs is mere speculation.

The third reason why thermodynamic stability is not per se a sufficient argument in favor of a predominant formation in the course of the CE is the fact that in a living organisms most biochemical reactions proceed with consumption of energy (e.g., ATP) under the influence of enzymes in a kinetically controlled reaction sequence and do not result in the thermodynamically most stable end product. For example, homochiral peptides and proteins do not represent a thermodynamically stable scenario, at least in solution. Here again an original comment of Luisi (p. 50):

> Is it the case that all compounds of our cellular life are thermodynamically stable? Of course, not [……]. Nowadays these compounds are formed thanks to the action of enzymes […..]. However, how could these compounds be formed in a time without enzymes? Do we need enzymes first, in order to make biochemicals that are not under thermodynamic control? [….] generally speaking this is one of the most difficult questions in the field of the origin of life: why and how have biological structures been formed that are spontaneous, i. e. not under thermodynamic control? […..] The notion of "frozen accident" is often used in the literature in this connection. This term conveys the idea that something that is not thermodynamically stable may have been formed by accident – and then been codified somehow in the living process. Of course, it is difficult to describe specific "frozen accidents" in terms of actual chemistry mechanisms. The notion of a "frozen accident" is a pictorial and fascinating metaphor, but it does not teach us anything from the operative point of view.

In other words and more clearly expressed, a fortunate event (coincidence) is needed to explain, what cannot be explained by rational, scientific arguments and experiments.

Concerning the question, why the canonical nucleobases of RNA and DNA were selected, the same problems arise as those discussed for prAAs. Why were the four nucleobases that form the basis of RNAs throughout the evolution over many millions of years once selected? In model experiments and in meteorites, nucleobases such as xanthine, hypoxanthine, and diaminopurine were found, which all are capable of base pairing. Furthermore, the author has not found any information about the potential prebiotic formation of analogs or purines such as amino and hydroxy benzimidazoles or benzopyrroles (Scheme 5.4). It is not clear, if it was impossible that those heterocycles, which can in principle compete with the formation of purine-based polynucleotides, were formed under prebiotic conditions, or if their formation in model experiments and their existence in meteorites was not explored.

Finally, the question needs to be answered, why D-ribose was selected as repeat unit of the RNA chains and not another chemically more stable saccharide. Addressing this question, Luisi summarized and cited the results of A. Eschenmoser and coworkers (p. 51). A. Eschenmoser and coworkers had synthesized several RNA analogs based on various saccharides, and he had compared the thermal stability of the double helices formed by those RNA analogs. He found, for example, for the double helix of a RNA analog prepared from D-ribopyranose a higher thermostability (Eschenmoser 1999; Pitsch et al. 2003). D-ribo-pyranose is a six-membered ring which involves the CH_2OH group for ring formation. He concluded that D-ribose was selected as standard component of all RNAs, because the lower stability of the double helices (incl. RNA/DNA helices) allowed for an easier separation of the strands and thus, more rapid, splitting into the single chains followed by copying at temperatures <50 °C. About several RNA analogs based on various pyranose rings, Eschenmoser said (Eschenmoser 2003):

> [....] these systems could not have acted as functional competitors of RNA in nature's choice of a genetic system, even though the six-carbon alternatives of RNA must have had a comparable chance of formation under conditions, where RNA was formed.

This line of argumentation has several shortcomings. First, it is in contradiction to Luisi's view that a high thermal stability of molecules (above all prAAs) was decisive for their selection in the early stages of the CE. Second, when various types of RNA type polymers were initially formed and were subject to a natural selection process, where did they know from that they had to develop a genetic code suited for the production of proteins? Without having this final goal in mind, the thermally most stable RNA analogs should have prevailed over the less stable ones, due to a higher resistance against hydrolysis and other degradation reactions. Why did the less stable normal RNAs have a chance to survive? This argument is aggravated by the assumption that locations with relatively high temperatures such as slopes of volcanoes or hot vents on the ground of oceans played a decisive role as birth places of the CE. When the CE has really started its career in the neighborhood of hot spots, Eschenmoser's argumentation leads exactly in the wrong direction.

Further experimental results with various monosaccharides let Eschenmoser later conclude that temporarily more effective replicators than RNA might have existed (Eschenmoser 2003):

While our experimental observations indicate that nature in selecting the molecular structure of the genetic system, had also other options besides RNA the notion we would naturally be inclined to consider, namely thatRNA might be the biologically fittest of them, al remains a conjecture.

Questions concerning the selection of bio-monomers, biopolymers, or secondary structures, may, of course, always answer that the CE has explored numerous alternatives and selected the fittest for life. However, with this "a posteriori" argument, one can justify the formation of all the monomeric, oligomeric, and polymeric molecules that are known from the living organisms today. With this argument in hand, part of the experimental research dealing with CE becomes obsolete. Furthermore, this kind of argumentation is in contradiction to the hypothesis of undirected CE and to the modern mainstream concept of evolution in the biosphere. The "a posteriori" argumentation sounds as if one says that the evolution of mammals had necessarily to result in the emergence of apes, because apes were needed to explain the appearance of humans at the (preliminary) end of the evolution. Hence, all the afore-discussed arguments only allow for one clear conclusion. On the background of an undirected evolution, no satisfactory answer can be given on the question "why this and not that," regardless whether monomeric, oligomeric, or polymeric species are considered.

11.2 Model Syntheses and Space Chemistry

Concerning the model syntheses of prAAs and in particular the syntheses described by the Miller group, a detailed critique has already been formulated at the end of Chap. 4 and only a few aspects of paramount importance should here be considered. Furthermore, critique of other model syntheses was added to Chaps. 5, 6, and 7, and thus, a detailed critique is again not needed at this point. Therefore, this section is mainly focused on arguments that were either not presented before or that need discussion in another context. Here, the the fact should be highlighted that almost all model experiments were performed with the intention to synthesize a certain compound or a special class of closely related compounds. When the syntheses were successful, the results were, of course, hailed as big success. However, the usefulness of such model syntheses as support for the hypothesis of a CE on the early Earth is rather low, if the reaction conditions are unfavorable for the formation of other classes of bio-monomers, bio-oligomers and biopolymers. When the reaction conditions are negative for the formation of other biomolecules, the success of the model syntheses is close to zero. This general comment should be concretized by the following examples.

First, the first experiment of Miller and coworkers (Tables 4.1 and 4.2) was conducted without H_2S or other sulfur containing gases in the reaction mixture. In this way, formation of Cys or Met was, of course, impossible. After Miller's death, former coworkers analyzed old experiments (stored in sealed glass flasks, Parker et al. 2011) by means of modern analytical methods, but neither Cys nor Met was detected. Furthermore, those authors did not disclose, if all the prAAs they had found in the absence of H_2S were formed again. Moreover, the formation of nucleobases was not studied. In other words, those experiments were a total flop.

Second, the first synthesis of adenine by J. Oro (1960) is another landmark in the history of model syntheses of biomolecules needed for a successful CE. However, the reaction conditions were absolutely unfavorable for the formation of pyrimidine bases and were also negative for the formation of almost all prAAs.

Third, the nucleoside imidazolides that were intensively used by various research groups, but mainly by L. Orgel and J. Szostack, as monomers for the preparation of numerous RNA sequences can easily react with the amino groups of amino acids and oligopeptides and hinder the formation of polypeptides and proteins. Furthermore, they react rapidly with the pendant amino groups of Orn, Lys, and Arg units and thus block the biological functions of proteins containing those prAAs. These "side reactions" also mean that RNAs were not accessible in the presence of prAAs and oligopeptides under slightly alkaline conditions as they exist in the human body and in the bodies of mammalians. This line of argumentation also concerns nucleotide triphosphates, because amino groups also react with any kind of phosphoric acid anhydrides.

Fourth, the experiments of M. Powner and J. D. Sutherland designed to prepare nucleotides without involving free D-ribose should be discussed (s. Sect. 5.4). Those authors have explored (and still explore) reaction conditions that enable the simultaneous synthesis of purine and pyrimidine nucleotides from a mixture of simple starting materials. A successful solution of this problem will, of course, be a big step forward for the RNA world hypothesis. However, the usefulness of the preparative success as support for the CE hypothesis still depends on the question, to what extent the reaction conditions are compatible with formation of prAAs, oligopeptides, and proteins. Powner and Sutherland need as starting materials, reactive electrophilic compounds, such as glycolaldehyde, cyanoacetylene, cyanamide, and diamino succinic dinitrile. These compounds can easily react with the amino groups of prAAs and peptides (incl. The pendant amino groups of Lys and Arg), and at slightly alkaline pH, they can react with the SH groups of Cys.

In other words, what is missing are experiments designed to compare to what reaction conditions that are favorable for the synthesis of type-X compounds are also compatible with the formation of type-Y molecules.

The last example already points to another group of problems, which are relevant for all model syntheses. The prebiotic syntheses of prAAs, saccharides, and nucleosides required a few simple but highly reactive starting materials, such as ammonia, hydrogen cyanide, and aldehydes, above all formaldehyde. Furthermore, several model syntheses of prAAs and the extracts of meteorites indicate that early oceans contained rather large fractions of carboxylic acids and alcohols. The term alcohol includes in this case di- and multifunctional compounds such as glycolic acid, glycolaldehyde, glycerol aldehyde, and dihydroxy-acetone. How was it possible that proteins and RNAs were formed in the presence of all these reactive molecules, which were badly needed, on the one hand, but also could play the role of trouble makers, on the other hand?

The following more detailed discussion of this problem is subdivided according to the four classes of "trouble makers" mentioned above.

Aldehydes

Formaldehyde is a component of interstellar clouds, and it is responsible for the formation of glycolaldehyde which has also been detected in interstellar clouds. Condensations of glycolaldehyde with formaldehyde or with itself yield higher saccharides ("Formose reaction"). Furthermore, glycolaldehyde is needed for the synthesis of purine and pyrimidine nucleotides via the "Powner-Sutherland chemistry" (Sect. 5.4). Moreover, formaldehyde is needed for the formation of glycine, and higher aldehydes are needed for syntheses of higher prAAs, such as Leu, Val, and Ile. In other words, formaldehyde is a basic building block for all three important classes of biopolymers. However, formaldehyde is a highly reactive electrophile which can react with any kind of amino groups forming aminals (hydroxymethyl amines) or Schiff bases. In this way, larger amounts of formaldehyde can prevent any synthesis of proteins. Hydroxy-methylation of only 5–10% of the pendant amino groups of Lys and Arg will destroy the biological function of all proteins containing these prAAs. Furthermore, <5% hydroxy-methylation of the amino groups of adenine or cytosine may suffice to affect base-pairing and biological function of RNAs. Therefore, larger amount of formaldehyde is deadly poison for all animals and in low concentration it causes allergies of humans by modification of proteins. Hence, it is justified to ask: Who has taught formaldehyde to present in sufficient concentration whenever it is needed for the synthesis of bio-monomers, but to be absent whenever proteins and RNAs? The author has never found any discussion of this problem in publication dealing with the hypothesis of CE.

Carboxylic Acids

Various types of carboxylic acids were discovered by Miller in his first studies of prAAs by means of electric discharges (Miller 1955). Most other research groups working on prebiotic syntheses of prAAs or nucleobases did not care about the detection of carboxylic acids in their reaction mixtures. For the question, if carboxylic acids might have hindered or at least disturbed the prebiotic syntheses of proteins, it is decisive that much larger amounts of carboxylic acids than prAAs were found in extracts of meteorites (s. Table 9.4). When a protein synthesis via activation of the carboxylic acid group is taken into account, carboxylic acids were unfavorable because they were activated in the same way and could react with amino groups. When the amino groups of prAAs or peptides were activated, the carboxylic acids have reacted with the activated amino groups. Hence, the assumption that protein syntheses were possible in primordial soups are nonsense and one has to take resort to the RNA world hypothesis and assume that RNAs were capable of developing a protein synthesizing machinery inside compartments that prevented an invasion of the interior by carboxylic acid. However, it will be hard to find a membrane which prevents penetration by polar carboxylic acids, such as formic acid or glycolic acid, but also prevents penetration by non-polar alkanoic acids. Anyway, only experiments can bring about a decision not arguments alone.

Alcohols
For several reasons, alcohols were certainly present on the early Earth in large quantities relative to prAAs or nucleosides. First, simple alkanols, such as methanol and ethanol, were detected in interstellar clouds and in extracts of meteorites. Extracts of meteorites also contained higher alkanols. Second, hydroxy acids, such as glycolic acid and lactic acid, were detected by Miller in his model syntheses of prAAs (Miller 1955), whereas other research groups did not care about formation of alcohols in their model experiments. Third, all condensation products of formaldehyde are (multifunctional) alcohols. Their presence or absence is relevant for the successful synthesis of RNAs, because any phosphorylation of nucleosides will also have involved other primary alcohols. The phosphorylated primary alcohols will then have, in turn, interfered with the polycondensation of nucleotides. When the formation of oligo and polynucleotides proceeded via polymerization of 2′,3′-nucleoside cyclo-phosphates, the negative influence of alcohol phosphates might have been negligible, but, once again the experimental evidence is missing.

Ammonia and Methylamine
Ammonia was used by numerous research groups as starting material for model syntheses of prAAs. However, ammonia is sensitive to UV light and its presence in the atmosphere of the early Earth is questionable. It is interesting to see that ammonia was not detected in interstellar clouds, whereas methyl amine and other amines were detected (Chap. 9). Methylamine was also detected in extracts of the Murchison meteorite. Furthermore, methyl isocyanate was observed in interstellar clouds, and its hydrolysis yields methylamine. Methylamine is for three reasons unfavorable for the formation of proteins and RNAs. First, it can compete with amino groups of prAAs and peptides in any synthesis of polypeptides and proteins. Second, it can cleave any phosphoric anhydride group or other activated phosphate group. Third, under condition yielding prAAs from gaseous starting materials, it yields sarcosine and higher N-methyl amino acids. This consequence is, for example, evident from the Miller experiments and from the extracts of meteorites. Incorporation of *N*-methyl amino acids into proteins destabilizes their secondary structure and hinders the desired biological function, even when only one Sar unit has been incorporated. How were all these problems avoided in the course of the CE?

When all these problems and arguments are taken serious, one can hardly escape from the conclusion that a successful course of the required a series or chemical miracles.

Finally, the results obtained from the examination of the space chemistry should be summarized. The most conspicuous negative result is the fact that numerous compounds that play a fundamental role for a successful CE were not found. Here, the following compounds should be mentioned: several prAAs such as Arg, His, and Cys, and oligopeptides; furthermore, the nucleobase cytosine, any kind of nucleosides, nucleotides, and oligonucleotides; last but not least, unsubstituted long alkanols or alkanoic acids which are needed for the formation of lipid membranes are extremely rare compared to the methyl substituted analogs. Even, when traces of some of these compounds will be detected in the future, such findings will not change the conclusion that the space chemistry was and is far from supporting a CE. This conclusion is confirmed by the fact that the space chemistry creates large amounts of chemicals that are unfavorable or even a definite obstacle for a CE. Such compounds are large amounts of non-prAAs and imino acids, large amounts of formic acid, alkanoic acids, hydroxy acids, and even sulfonic acids. The space chemistry is characterized by an extreme diversity driven by a gain of entropy. The numerous aggressive radicals and ions that were discovered in interstellar and circumstellar clouds fuel the chemical machinery which creates the extreme molecular diversity. Nonetheless, it makes sense to search for life on other planets, simply because life exists on Earth. Consider, living organisms regardless, how primitive or intelligent, will be found somewhere else in our galaxy, what does it mean in terms of fundamental knowledge. Such a discovery tells us that a special law of nature exists, which says that under favorable conditions, life will emerge, although not as direct consequence of space chemistry. This law will not be an alternative or contradiction to the known laws of chemistry and physics, it must be understood as a complementary, overriding principle. Without knowledge about life on other planets, the conclusion of the chemist D. Rehder is justified, who said in his book "Chemistry in Space" (Rehder 2010) that the emergence of life on Earth is the result of an extremely unlikely chance event.

11.3 The Primordial Soup and Other Fundamental Hypotheses

For a correct and fair evaluation of the primordial soup hypotheses of A. Oparin and J. Haldane, one has to distinguish between their historic merits, on the one hand, and their usefulness and importance as evidence for the CE, on the other hand. The historic merits are beyond any doubt and don't need any discussion. The impact of those hypotheses on theoretical and experimental activities concerning the origin of life has much to do with the mental and theoretical vacuum in this field following the discoveries of L. Pasteur in the 1860s and 1870s. The view of Aristotle that primitive organisms can emerge from dust and garbage at any time (provided the temperature was high enough and moisture was present) was the prevailing theory for almost 2000 years. L. Pasteur's experiments unambiguously demonstrated that Aristotle's view was wrong; living organisms always have or had living parents regardless of their primitivity. Yet, Pasteur did not forward an alternative theory explaining origin and existence of living organisms. Hence, there was a mental vacuum concerning the understanding of life's origin in the five decades following Pasteur's experiments. Oparin's and Haldane's primordial soup hypotheses presented an entirely new approach to a scientific understanding of the origin of life. Their hypothesis stimulated the development of modified and more detailed versions of their view, and they opened the door to experimental studies.

These immense historic merits of Oparin's and Haldane's work stand now in sharp contrast to what can be said today about the usefulness as birthplace of a CE yielding the first living cell. Certainly, the first oceans have accumulated organic matter from two sources. There was an influx of organic compounds from space via meteorites and above all via micrometeorites. Furthermore, organic compounds were synthesized on Earth from gases such as CO_2, CO, N_2, and H_2 in combination with water. In addition to the hot vent hypotheses already discussed in Chap. 2, new results of the Trapp and coworkers should here be mentioned (Peters et al. 2023). This research group has quite recently demonstrated that iron particles coming in from meteorites or being present on the Earth have played an important role. They reduced water at high temperature and under pressure, and the resulting H_2 reduced CO_2 and N_2. Iron itself and traces of other metals in combination with minerals (e.g., clay) played the role of catalysts. The accumulation of organic matter in early oceans is, of course, a prerequisite for a chemical evolution, but for reasons discussed above not any compilation of organic matter is suitable to initiate a CE in the primordial ocean. Critique of the primordial soup theory of Oparin and Haldane began to appear in the 70s of the last century. For example, the geologists Brook and Sharp (1978) reported that in the oldest sediments and rocks that were studied at that time (3.5–4.0 million years old), organic matter was rarely found. Furthermore, the analysis of those organic compounds revealed that they almost exclusively consisted of hydrocarbons and contained only traces of nitrogen (<0.2%). These results are incompatible with a primordial soup containing significant amounts of cyanamide, amino acids, and nucleobases. Further critique was, of course, uttered from all those scientists who proposed another birth place of life. The chemist Cairns-Smith (1982, s. Sect. 2.4), the astronomer F. Hoyle (Hoyle and Wickramasinghe 1981), the biologist C. Woese (1980), and the biochemist R. Shapiro (1987) belong to this group. Shapiro wrote:

> The Oparin thesis has long ceased to be a productive paradigm. It no longer generates novel approaches to the problem; more often than not it requires modification to account for the facts, and its overall effect now is to stultify and to generate disinterest in the problem of life's origin. These symptoms suggest a paradigm whose course is run, and that is no longer a valid model of the state of affairs.

These and other critical arguments were summarized by R. Shapiro in his book "Origins" in two Sections entitled "The myth of the prebiotic soup" and "The retreat from the hypothesis." However, it is the view of the author that four more arguments against the primordial soup hypothesis must be considered.

First: As discussed above, the presence of significant amounts of aggressive chemicals, such as formaldehyde and ammonia, which are needed for syntheses of prAAs, saccharides, and nucleobases prevent formation of RNAs and proteins.

Second: How was it possible that in a mixture of prAAs with an excess of numerous non-prAAs and imino acids a selective polymerization of prAAs has taken place? Analogously: how was it possible that in a mixture of various N-heterocycles and saccharides, a selective synthesis of certain nucleosides and RNAs has taken place?

Third: According to the modern theory of step-growth polymerizations (Chap. 3), polcondensations of diluted monomers primarily yield cyclic oligomers and certainly not high molecular weight polymers.

Fourth: At the high temperatures postulated by Haldane, racemization of prAAs proceeds quite rapidly, so that formation of homochiral proteins is absolutely unlikely.

Today one has to characterize Haldane's hypothesis that a "hot dilute soup" was the birth place of the CE not only as unlikely but as non-scientific. However, the successful biologist and genetician J. Haldane cannot be blamed for the fundamental defects of his hypothesis, because at the time when he wrote his book, a polycondensation theory did not exist yet, and all the information and experimental facts known today were not predictable. All the critical arguments summarized above are also valid for Oparin's first concept of a primordial soup. However, in contrast to J. Haldane, Oparin has modified his first hypothesis in the course of the three decades following his first publication, and the concept that coacervates played an important role as "mini-reactors" gave his hypothesis a more modern more convincing touch. Unfortunately, Oparin postulated for the formation of coacervates the availability of biopolymers in the primordial soup, but even in 2023, no experiments and hypotheses exist explaining how these biopolymers might have been formed in a primordial soup. Furthermore, Oparin did not postulate and discuss that coacervates need semipermeable membranes when acting as reactors for a CE in an environment that was more or less different from the primordial soup. Therefore, the afore-mentioned critique of C. Woese is still valid without restrictions.

On the other hand, one can understand Oparin's concept of coacervates as precursor of modern compartment theories and thus, as precursor of the RNA world hypothesis. Solely in compartments and in by interaction between monomers and oligomers with the inner surface of compartments, it is imaginable that polycondensation reactions have successfully proceeded. It may also be hypothesized that cooperation between two different types of compartments, namely inorganic compartments such as silicates or ice, and organic compartments cased on lipid layers has played a role for limited period of time. The enormous progress which has been achieved within the framework of the "evolution in a test tube" (Sect. 8.2) renders the combination of RNA world hypothesis with the compartment hypothesis, the only plausible candidate for a scientifically acceptable concept of an explanation, how a chemical evolution might have taken place. However, numerous critical questions are still open as discussed in detail in Chaps. 2, 3, 4, 5, 6, 7, 8, 9, and 10 and those critical comments should not be repeated again.

Nonetheless, it is amazing that even after the year 2000, the prebiotic soup hypothesis still plays a role as standard scenario for a CE, which yielded a living cell starting out from a mixture of simple molecules such as CO_2, N_2, H_2, and H_2O. This view is above all characteristic for biologists as may be demonstrated by a comment of Ribas de Pouplana in his book "The genetic Code and the Origin of Life" (Ribas de Poublana 2004):

> The most widely accepted scenario for the transition from abiotic to biotic chemistry is that the simple monomeric compounds present in the prebiotic soup somehow underwent polymerization, perhaps with the assistance of clays and minerals and formed longer and longer chains or polymers, which over time became increasingly more complex with respect to both their structure and their properties.

In this way, the wonder of the primordial soup as the birth place of life will be preserved for eternity, at least for laymen of chemistry.

Finally, the fundamental problem should be discussed, namely the question, why a CE on the early Earth has resulted in the formation of a living cell. Even, when the number of model reactions demonstrating how individual steps of the CE might have occurred will further increase, the knowledge of hundreds of individual reactions does not necessarily explain, why a CE followed the road map toward a living cell. Increasing complexity of groups of molecules also means increasing diversity for the entire pool of organic compounds on Earth, but the driving force for an increasing diversity is a gain of entropy. This correlation is best and clearly demonstrated by the "space chemistry." However, a living cell practices a different strategy, it maintains thermodynamically instable, kinetically controlled processes and compounds against a gain of entropy by permanent consumption of (sun) energy.

The hypothesis of a fortunate coincidence or contingency is for three reasons not an acceptable hypothesis of CE. First, scientific research concerns discovery evaluation and application of laws of nature. A law of nature is evident from the reproducibility of (experimental) observations which are reproducible independently of time, location, and researcher. Yet, singular events are per definition not part of a law of nature, and thus outside the reach of scientific methods. Second, it has not been demonstrated yet that all the components needed for the formation of a living were simultaneously present on the early Earth, regardless whether the primordial soup or a compartment concept is taken into account. The experiments and observation listed in the preceding chapters rather suggest that a more or less simultaneous evolution of homochiral proteins, RNAs, and amphiphilic molecules forming stable vesicles was not possible. When the RNA world hypothesis is taken into account and when an independent evolution of proteins is ignored, one still needs a strong belief in chemical miracles to assume that those RNAs needed for the production of bioactive proteins were formed in significant quantities, came together and recognized that they had to cooperate for the production of useful proteins. Third, it requires again a strong belief in wonders to expect that the problem of homochirality was solved by a chance event. Even when one concedes that a fortunate event has accumulated homochiral biopolymers, the racemization would have destroyed the homochirality within a short time except at temperature around or below 0 °C.

11.4 Did Life Emerge Only Once in the Universe?

This question cannot be answered at this time, and only the extensive and time-consuming exploration of exoplanets might give a clear answer on this question. However, the mental prerequisites needed for correct answer on this question may defined, and thus, avoid misunderstandings and misinterpretations. A correct response can only be obtained on the basis of the following alternative concepts.

I. The emergence of life is a consequence of one or more laws of nature. In this case, the emergence of life must have taken place million or billion times in the universe and will continue million or billion times in the future considering the existence of billion of galaxies. However, the timely and spatial distance between these events may be so huge that neither the humankind nor any other "civilization" of living organisms has a chance to detect the existence of life on another planet.

II. There is no law of nature responsible for a reproducible emergence of life. In this case, the life on Earth is a singular event in the entire universe.

Ad I. If life has emerged and will emerge on numerous planets of the universe, the law of nature governing these CEs is a special law which is complementary to the existing laws of chemistry and physics and not in contradiction to them. It may be understood as an overriding trend which becomes evident as soon as the conditions on the planet meet the necessary "reaction conditions." As soon as life has been found on an exoplanet, it is clear that a law of nature favoring the emergence of life exists in the universe. But let's consider that the scientists were not able to detect life on 100 exoplanets during the next 300 years. What is the correct interpretation of this negative result? Is hypothesis II the unavoidable answer? It might be that the timely and spatial distance between these events may be so huge that neither the humankind nor any other "civilization" of living organisms has a chance to detect the existence of life on another planet. In summary, even when the humankind does detect life on exoplanets, we will not have definite answer if hypothesis I or II is correct.

Ad II. If the emergence of life on Earth was and is a singular event in the universe, scientific methods cannot answer the question why, in close analogy to the question what is the origin of the "Big Bang." However, scientific research may, in the optimum case, explain the course of the CE in terms of individual reaction steps and reaction sequences. Two scenarios may be distinguished. Either, the appearance of the first cell was sudden event. In this case, a divine creational act is one potential explanation. The alternative hypothesis is the assumption that the properties of the universe include a sudden appearance of a living cell as a kind of second "Big Bang." As already mentioned above, neither explanation allows for scientific studies.

11.5 Final Comment

The reader would presumably like to know what is final conclusion of the author. In short, the author is not able to reach a final decision. The hypothesis of a CE is a fascinating vision which (as pointed out below) allows for a consistent interpretation of the structure of the universe, but there are so many open questions and unsolved fundamental problems that the author is not convinced yet that a CE beginning with simple gases and ending up with a living cell has really taken place. A hypothesis is a preliminary explanation of experiments observed in laboratories or of observations made in nature. New experiments stimulated by the hypothesis may have three consequences: first, they confirm the hypothesis, second, they require (slight) modifications, or third, they prove the hypothesis being wrong, so that a new hypothesis is needed. A scientist has to be open to all three consequences, and the author is willing to accept any of these consequences for the hypothesis of CE. A researcher, who is a priori convinced that the hypothesis of CE is true, he takes an ideological stance and not a scientific one.

Undirected evolution in combination with chance events is not a scientific approach to the explanation of a CE (because reproducibility is fundamentally impossible) and requires so much believe in miracles as any religion. The hypothesis of a directed evolution is confronted with the question who was the "director,", or which property of the universe has taught the molecules to follow the pathway to a living cell. This problem is acerbated by the finding that the reductionistic-deterministic concept is not correct as evidenced by the failure of numerous model experiments and by the course of the "space chemistry." One potential answer is the religious one, meaning that a God (Allah, Manitou, etc.) has directed the molecules. The author proposes two non-religious alternatives. Firstly, one may simply assume that the design of the universe includes a directing power, which is operating, whenever the reaction conditions on a planet enable a CE. Secondly, one may say that the design of the universe is based on six dimensions: space, time, life, and the mental activities of the humankind (philosophy, art, science). This concept has three merits. It is a consistent interpretation of all we know about the universe and our life, it is valid regardless if life emerged once or many times in the universe, and it allows to explain why the humankind might be the goal of the evolution. With the brain of the humankind, nature has realized a scenario where it can study itself or communicate with itself. This interpretation does not necessarily mean that the humankind on Earth is unique in the universe. The humankind may vanish and nature initiates (and already has initiated) analogous evolutionary processes with the same goal somewhere else in the universe. When the combined chemical and biological evolution up to living organisms that can communicate with nature followed by their rapid or slow disappearance happened and happens many times in the universe, this phenomenon may be seen as analogy to birth, growth, and death of stars.

When the existence of a CE is denied, again a religious and a non-religious answer is possible. The religious answer is the creation of the first cell by a God (Allah, Manitou, etc.). The non-religious view is to say that the design of the universe includes two "Big Bangs," the first one resulting in the appearance of space, time energy and matter, and a second "Big bang" resulting in the appearance of life with the first cell. Any non-religious hypothesis is confronted with the open question about the origin of the first "Big Bang," but this point does not concern the hypothesis of CE. In summary, the author is inclined to accept the previous existence of a directed CE, but this hypothesis still lacks sufficient experimental evidence for being a convincing theory.

References

Brook T, Sharp G (1978) Orig Life 9:297
Cairns-Smith A (1982) Genetic take-over. Cambridge Univ. Press, Cambridge
Eschenmoser A (1999) Science 284:2128

Eschenmoser A (2003) Creating a perspective for comparing. In: Proceedings of the J. Templeton Foundation "biochemistry and fine-tuning", Oct. 10–12. Harvard Univ. Press

Hoyle F, Wickramasinghe C (1981) Evolution aus dem All (evolution from space). Ullstein, Berlin

Luisi P (2006) The emergence of life. Cambridge Univ. Press, Cambridge

Miller S (1955) J Am Chem Soc 77:2351

Oro J (1960) Biochem Biophys Res Commun 2:407

Parker E, Cleaves H, Callahan N, Dworkin J, Glavin D, Lazkano A, Bada J (2011) Orig Life Evol Biosph 41:201, 569

Peters S, Semenov D, Hochleitner R, Trapp O (2023) Nat Commun Sci Rep 13:6843

Pitsch S, Wendekorn S, Krishmanurthy R, Holzner A, Minton M, Bolli M, Mikulka C, Windhab N, Micuru R, Stanek M, Eschenmoser A et al (2003) Helv Chim Acta 86:4270

Rehder D (2010) Chemistry in space. Wiley VCH, Weinheim

Ribas de Poublana L (2004) The genetic code and the origin of life. Kluwer Academic and Plenum Publ., New York

Shapiro R (1987) Origins – a sceptic's guide to the creation of life on earth. Bentham Ed, Summit Books, a Division of Simon & Schuster, New York

Wills C, Bada J (2000) The spark of life. Perseus publications, Cambridge

Woese C (1980) An alternative to the Oparin view of the primeval sequence. In: Halvorson H, Van Holde K (eds) The origin of life and evolution. Alan R. Liss, New York, pp 65–76

MIX
Papier aus verantwortungsvollen Quellen
Paper from responsible sources
FSC® C105338

If you have any concerns about our products,
you can contact us on
ProductSafety@springernature.com

In case Publisher is established outside the EU,
the EU authorized representative is:
Springer Nature Customer Service Center GmbH
Europaplatz 3, 69115 Heidelberg, Germany

Printed by Libri Plureos GmbH
in Hamburg, Germany